PLANTS AND PLANTING ON LANDSCAPE SITES

Selection and Supervision

Plants and Planting on Landscape Sites
Selection and Supervision

Peter R. Thoday
Former Senior Lecturer in Amenity Horticulture
University of Bath, UK
&
Principal of Thoday Associates, Landscape Consultants, UK

CABI is a trading name of CAB International

CABI	CABI
Nosworthy Way	745 Atlantic Avenue
Wallingford	8th Floor
Oxfordshire OX10 8DE	Boston, MA 02111
UK	USA
Tel: +44 (0)1491 832111	Tel: +1 (617)682-9015
Fax: +44 (0)1491 833508	E-mail: cabi-nao@cabi.org
E-mail: info@cabi.org	
Website: www.cabi.org	

A catalogue record for this book is available from the British Library, London, UK.

Library of Congress Cataloging-in-Publication Data

Names: Thoday, P. R., author.
Title: Plants and planting on landscape sites : a guide to plant selection and site supervision / Peter Ralph Thoday.
Description: Boston, MA : CAB International, [2016] | Includes bibliographical references and index.
Identifiers: LCCN 2016013392| ISBN 9781780646183 (hardback : alk. paper) | ISBN 9781780646190 (pbk. : alk. paper)
Subjects: LCSH: Landscape design. | Landscape plants.
Classification: LCC SB472.45 .T46 2016 | DDC 712--dc23 LC record available at https://lccn.loc.gov/2016013392

ISBN-13: 978 1 78064 618 3 (hbk)
          978 1 78064 619 0 (pbk)

Commissioning editor: Rachael Russell
Editorial assistant: Alexandra Lainsbury
Production editor: James Bishop

Typeset by SPi, Pondicherry, India.
Printed and bound in the UK by CPI Group (UK) Ltd, Croydon, CR0 4YY.

For Mark Hendy without whom there would be no books by Peter Thoday

Contents

Foreword

Perhaps more than any other design discipline, landscape architecture combines the arts and sciences, incorporating aspects of botany, horticulture, the fine arts, architecture, planning, soil science, geology and ecology. Not only are good landscapes defined by their initial appearance, they are also judged on how they develop over time, requiring an understanding of how plants establish and how they change as they mature. This requires a degree of confidence that plants will grow and flourish in accordance with design intent, which can only be achieved in practice through the application of scientific principles. Without a sound understanding of the natural world and the processes that enable plants to grow and flourish the most artistic planting designs will ultimately disappoint.

With an academic background firmly rooted in amenity land management and horticulture, developed over more than 25 years as Senior Lecturer in Amenity Land Management and subsequently Director of Studies for Horticulture at the University of Bath, Peter Thoday has dedicated a significant part of his professional life to understanding amenity planting, leading to appointments as Horticultural Director to the Eden Project, Cornwall, and as Chairman of Trustees for the Sensory Trust, promoting access for all to public open space and the countryside.

In his role as horticultural consultant and Principal of Thoday Associates, his pragmatic approach to solving the many and varied practical problems encountered on development sites and in the restoration of historic landscapes has informed the successful establishment and long-term success of many landscapes throughout the UK and abroad helping design professionals at all levels to achieve their design intent through the application of sound horticultural principles.

Fundamental to this approach has been an appreciation that, while all sites differ in terms of micro-climate and soils, and plant types may differ from each other in terms of their specific requirements for healthy growth, the same basic principles of plant establishment apply, regardless of whether a scheme involves small-scale domestic style planting with predominantly herbaceous material or the creation of a large-scale commercial landscape comprising mostly woody plants and shrubs.

Design professionals and contractors alike are required to have a good working knowledge of how to achieve plant establishment under a variety of conditions and situations. Standard planting specifications with alternative clauses are available and are of immense value to designers in this respect, but are only truly of use if there is an understanding of the science behind each clause, enabling informed decisions to be made by the designer in choosing clauses that are directly relevant to a particular project and discarding those which are not.

This book is therefore of value to anyone involved with the writing of planting specifications and the design and implementation of amenity planting schemes, from students seeking a sound understanding of how to source and establish plant material, to professional landscape architects, garden designers, horticulturalists and specialist landscape contractors requiring a checklist or quick reference of what needs to be considered when approaching a new project. It may be read as an informative textbook giving an overview of the topic or referred to section by section as needed. Where related content falls outside the scope of the book further reading is suggested by the author.

Throughout the book there is an understanding that achieving best practice in plant establishment requires an appreciation of plants as living things and application of technical knowledge at every stage in the design process and life cycle of a project, from initial site surveys, through site planning, ground preparation and planting, to maintenance and management operations.

As a landscape architect, I am immensely grateful to Peter for his unwavering support and advice over many years and I have come to view him not only as a reliable and trustworthy dispenser of sound technical advice, but also as a mentor and friend. I know that I am not alone in appreciating these traits and I have no hesitation in thoroughly commending this book in the expectation that many others will benefit from his extensive knowledge.

Colin Brown BSc (Hons) Dip LD MA FLI

Preface

Topsoil shifting, sourcing and transplanting nursery stock and arranging for irrigation all sound familiar tasks but this particular list is not taken from any recent project. They are transcriptions by Assyriologist Professor Stephanie Dalley of the clay tablets written by King Sennacherib in 689 BC describing the making of the stupendous grounds of his great place at Nineveh now thought to be the real *Hanging Gardens of Babylon*.

Over the succeeding millennia, planters have faced the same challenges as they planted the surrounds of their civic buildings and public spaces. History has left us a handful of illustrations of mature medieval landscapes along with the unique master plan of the monastery of St Gaul dated 820 AD, which went as far as to name the vegetables in the ascribed beds. It is not until the modern era that we learn the names of designers and it is even later before anyone who carried out such work is identified. Throughout this long history it is thought likely that the planting was directed and executed in house with no attempt to finish concurrent with the completion of the building.

Almost nowhere in the extensive pre-20th century horticultural literature do we find any reference to the risk of losses or poor growth following transplanting. Strangely well-founded concerns arise in the era of huge advances in the plant sciences and agronomy and after the professionalization of landscape design and the establishment of numerous courses in amenity horticulture.

How has this happened?

Some problems seem to result from plant science and husbandry being short-changed in the education of both designers and planters. This can result in poor specification and plant selection or poor work on site; however, in many cases the root of these problems lies elsewhere.

Both landscape design and planting are frequently incorporated into the project's overall management masterplan and construction programme. This apparently eminently sensible move brings with it certain risks.

In some business models such as design and build, the designer's input may be terminated following the acceptance of the detailed design, thereby removing his/her role in supervising work on site with their unique understanding of the design intent.

A very different and truly down to earth problem that results in years of poor slow growth occurs when ground works, including soil moving and storage, no longer involve the horticultural team. With no direct interest in the damage to the soil that their powerful machines can cause, ground-work contractors may tackle most tasks in almost all weathers. This results in damage to soil structure, compaction and impeded drainage. With time such damage can be rectified but in many large projects the horticultural work must fit into the overall programme and be finished by the conclusion of the building contract. This can lead to land preparation and planting being done under inappropriate conditions.

Plantings are envisaged by designer, client and planning authority in their maturity to provide such functional roles as screening, shelter, scale and shade so it must be in the interest of all parties to achieve the desired degree of maturity as soon as possible. Why then cripple plantings with years of poor, slow growth that results in a failure to achieve the design intent in order to have an 'instant' but inevitably temporary effect on commissioning day? This desire to make an immediate impact lies behind the use of inappropriately sized transplants for the care they will receive and too high plant density for long-term sustainability.

While this Chelsea Show garden approach lies behind some long-term disappointments, in many cases a shortage of money is at the root of the problem. Landscape budgets tend to be cut if there is an overall project overspend but even without this both designers and planters operate on very tight margins to a point where the former can make only the most critically important site visits and the latter cannot afford the time needed for the skilled hand work required to rectify adverse condition and / or maintain plantings during the defects liability period.

The aim of this book is neither to harp back to long gone traditions nor to unfairly criticize the various groups from designer to stock producer to those who do the site preparation and planting; nor is it to produce a didactic landscape works formula that looks good on paper but is unworkable on site. The hope is to help provide a fuller insight into the several stages that must follow one another if the design intent is to be achieved through healthy plantings. Amenity horticulture is poorly researched compared with the crops of agriculture, horticulture and forestry. In the absence of such 'hard quantified data' I have, wherever possible, linked site observation to plant biology: our understanding of the growth of plants and their genetically based response to conditions in their immediate surroundings. In spite of the range of species and site conditions involved I consider this remains the best approach to amenity plant husbandry. Only by making such a link can the difference between unnecessary tasks and critical inputs be judged and the skills within the horticultural team put to best use. All the problems I have flagged up have been encountered several times during a long working life, though not all at the same time or all on the same site!

By far the most pleasurable part of writing a preface to one's book is the opportunity it provides for the writer to thank the people that have given time, advice, information and, above all, encouragement. But before thanking these kind folk I must take responsibility for my text and any errors it may contain.

Against the convention of putting ones family last I start with my greatest helper in every way, my wife Anne. My colleague in Thoday Associates Mary Payne has generously shared her experiences in the field, provided editorial comments and contributed many of the photographs.

Colin Brown has written a thoughtful, kind and supportive foreword and provided an examiner's understanding of the needs of new entrants to landscape architecture.

Ben Kirby's clear drawings, spread throughout the book, are certainly worth several thousand words.

As with previous books Mark Hendy gave my often-muddled original jottings form and structure enabling me to present a coherent submission to publishers CABI.

Richmond Dutton provided much of the information on brownfield sites and the loan of his PhD thesis.

A combination of Landlife's publications, photographs and conversations with Richard Scott were invaluable when introducing the use of native species.

I am grateful for information and additional comments on tree root systems from Peter Gregory at East Malling Research.

I extend my thanks to my very long standing friend Mike Wilson for having nurtured the avenue we planted together 16 years ago, prior to his wife Julie Yager taking the photograph reproduced on the cover.

Thanks go to the following friends whose permission to use their photographs so enrich this book: Mary Payne, Andrew McIndoe, Richard Scott, Tony Kendle, James Hitchmough, Mikel Pagola and Forest Research.

IT expert Jacob Payne has once again sorted out both hardware and software computer problems and valiantly attempted to teach us new skills.

My thanks go to publishers CABI. While renowned worldwide for their work in supporting crop producers they may not be so well known among landscape architects and amenity horticulturalists as *Plants and Planting on Landscape Sites* is a first venture into this area. I particularly wish to thank Commissioning Editor Rachael Russell for her support and guidance and freelance editor Frances East for her skill and the tact she showed in sorting out my omissions. Production editor James Bishop heads my final list of well-deserved thanks. Publishing is a multitask operation and I am grateful to those who are sometimes overlooked, the proof reader, typesetter, index compiler and the printers; without their combined skills there would be no book.

Peter Thoday

Glossary

Apical dominance The control of the pattern of growth exerted by the terminal bud

Broadcasting The distributing of seeds by hand onto the soil surface

Burlap A course weave fibre used to contain roots with attached soil

Calcicole A plant that flourishes in chalky, limestone rich soils in cultivation often equated with neutral or slightly alkaline soils

Calcifuge A plant that in nature grows in lime-deficient soils. In cultivation, this equates with needing acid soil

Cambium Cells found within stems and roots that can divide to produce new tissue and thereby organs

Capital works All the tasks involved in the execution of a project before coming under the management of the client

Capping The forming of a surface crust of denatured soil. Often detrimental to the passage of air and water and the germination of seeds

Check A physiological condition brought about by environmental conditions in which growth is greatly reduced

Coppice The act of, or the resultant growth from the cutting down of woody plants

Danish trolley A hand-propelled truck designed and used for transporting containerized plants

Etiolated growth Stems and leaves produced under low light, typically having long, thin, pale internodes between leaves

Field capacity The maximum amount of water that a freely draining soil holds against the pull of gravity

Forbs Species other than grasses growing in a grassland sward

French drain A clean aggregate-filled drainage ditch generally used to intercept surface runoff

Genotype The genetic makeup of a plant formed from the two sets of chromosomes (the genome) inherited one from each parent

Gross morphology The overall shape of a plant resulting from either its nature or nurture

Hardening off The gradual exposure of a glasshouse or poly-tunnel grown specimen to outdoor conditions. Also termed acclamation, it triggers significant physiological processes

Humus Partly decomposed organic matter that plays a vital role in soil fertility

Husbandry All the work and processes concerned with the cultivation of plants

Hydroseeding The application of seeds from a spray of water carrying suspended, often organic solids that form a mulch

Juvenile growth Shoots produced by a young plant while in its pre-reproductive stage. Similar growth can result from severe pruning

Liner A small container-grown transplant in the early stage of post-propagating production

Loamless compost A root substrate (potting compost) made from organic materials without the addition of any mineral soil

Mulch As referred to in this book, course chipped fragments of woody material spread on the soil surface to inhibit weed growth and retain soil moisture

Peds The naturally occurring soil aggregations of a well-structured soil between which water, air and roots pass

Pleached As now used the training and pruning of trees to produce an artificial geometric canopy

Radiation frost A period when the night temperature falls below freezing due to heat loss to clear skies; most damaging in spring when new growth has started

Ramet As used in this book, a fragment of a plant capable of producing a new individual

Root plate The shallow disk of roots radiating out from the butt of many trees, clearly evident on wind-blown specimens

Root zone The area of land occupied by the root system of a plant

Seed bank The number and range of live seeds held in the soil of a given area

Stolons Long, thin, more or less horizontal or trailing shoots from which plants develop at the nodes

Swale A ditch usually constructed to carry water only during storms or throughout wet periods

Sward A grass cover either with or without the presence of other species

Taproot The root developed from the radical or first root held within the seed. Also used more loosely to describe a root or roots growing down through the soil profile

Taxa (singular taxon) A term describing any grouping of individuals regardless of their status within a taxonomic system

Terminal bud The bud at the tip of a stem. Through the production of growth hormones it controls the development of shoots lower down the stem

Undercut The severing of roots without lifting the plants by using a blade drawn through the soil

Water table The level below which the spaces between the mineral particles are filled with water. The position of a water table typically fluctuates with the seasons

Introduction

1

Plants play a vital part in the ambience of the vast majority of landscape schemes, ranging from roads to business parks, schools and hospitals. The 'green component' is, at times, seen by developers primarily as a planning requirement but by the public as contributing to the attractiveness of the scheme, a quality future occupiers hope will enhance their image.

In spite of such approval, there remain many examples of poor development site plantings that fail to achieve their potential, many of them set within excellent overall architectural and landscape designs. Failure in this context very rarely means the death of all the plants. It is usually either the death or very poor condition of sufficient individual plants to ruin a planting's composition, or that growth is so compromised that it takes years, if ever, for the specimen to achieve the size and form intended. Richmond Dutton (1991) in his research at Liverpool University found that while a few such failures were due to poor design detail, and rather more to inappropriate plant selection, the majority resulted from poor work at either the capital works or the maintenance stage. Clearly, gross malpractice leads to failure but this is rarely the cause; usually it is the omission or poor execution of husbandry tasks that, while apparently very simple, are in fact quite critical. Often this can be traced not to the ignorance or incompetence of the workers on site but to horticultural work and, in particular, soil handling being attempted in conditions that no one in any other branch of agronomy would contemplate working under. These circumstances are usually attributable to the dictates of the construction programme.

Concerns and responsibilities for the success of plantings start long before work on site and, as one senior landscape architect put it, can even include 'the need to communicate what plants 'are' (i.e. that they are alive) to others within a design team'. It is equally important to ensure that the client understands what can be expected from the plantings – in terms both of their appearance at the completion of capital works and of how they will grow towards achieving the final design intent.

This book addresses those matters that arise during the preparation and planting of landscape sites. The success of such undertakings rests upon the creativity of the designer combining with the craftsmanship of the contractor. Both contributions are essential, and while the contractor is bound by the terms of the contract and the designer, in the guise of the supervising officer, is committed to act on behalf of the client, they share the common objective of creating a successful landscape. Within that challenge falls the selection and quality of the plant material, the site and its preparation, the planting operation and the management of the site during the establishment or 'defects liability' phase. It follows that when dealing with the soft landscape, those areas of soil within the design, designers must be sufficiently knowledgeable to recognize the advantages and limitations of the site's immutable features together with both good-quality (and bad-quality) plants and good and bad horticultural practice.

In these pages an attempt is made to present a balance between providing some background insight through reference to plant and soil science and more directly applicable information on aspects of plant husbandry. In some cases decisions must made on the basis of local conditions and may even run contrary to what would otherwise be best practice; nevertheless it should be remembered that the fundamental aspects of plant biology remain both in play and unchanged.

Creating a living landscape requires at times a multidisciplinary approach, calling on the expertise, whether field or laboratory based, of, among others, drainage engineers, soil scientists, horticulturists and foresters. The more challenging the site the more such experts should be involved. Plantings are unique in that they are not, and cannot be, at their best when handed over to the client. A planting, unlike a hard landscape, takes months or years to deliver the designed intent. On completion of the capital works the responsibility to achieve and sustain that development passes to those who manage and maintain the planting.

Plants as Living Organisms

Basic requirements

Before considering the specific qualities required of landscape plants (see Chapter 5) we should recognize those characteristics that all plants share and that set them apart from all the other materials used in constructing a landscape or garden. To most readers it will seem unnecessary to point out that plants are living organisms and must be treated as such. It is not unknown, however, to find plants on site, both before and after planting, treated rather as if they were a stack of bricks; indeed the perishability of bags of cement may receive more understanding and care.

In common with all other life forms, plants require a number of conditions to support their basic metabolic processes and hence keep them alive, namely sufficient heat, light, water and nutrients. A plant's tolerance of a shortage of such requirements may be more or less critical and varies from species to species.

- Heat: All plants require heat for basic metabolic processes but the level of high- and low-temperature tolerance varies greatly between species, producing various degrees of hardiness as discussed in the Hardiness section, page 48.
- Light: Light is the source of energy for photosynthesis upon which all green plants depend; however, there are big differences between the needs of shade-tolerant and sun-demanding species. Also, planting site light levels may be influenced by the shade of buildings and overhanging vegetation, a topic considered more fully in the Shade tolerance section, page 49.
- Moisture: Plant tissues dry out when water loss by evapo-transpiration exceeds water uptake. Impeded uptake occurs as a result of any one of the following: the root system is insufficient or dead; roots are exposed to the air; or the soil within the root zone is dry or not in contact with the roots. The application of water by irrigation is discussed in Irrigation at and after planting section, page 129.
- Nutrition: Although the availability of the elements required for plant growth is essential in the long term, their level is unlikely to influence transplant establishment. The timing of any additions will depend on the fertility of the site soil and the season in which transplanting takes place (see page 16, section on Soil fertility).

Requirement or tolerance?

Most cultivated plants have retained many of their evolved ecological responses to conditions that prevail in their natural habitats. Each environmental factor has an optimum level for the plant, at least in theory, although plants can tolerate some excess or deficiency albeit with reduced performance. Because few sites provide optimum conditions for a wide range of plants, the selection has to be made from those plants whose tolerances to such factors as water, light, soil pH or salinity fall within the ranges of local levels.

There is often a confusion between a plant's essential requirement for various environmental factors and its tolerance of them. We may refer to a plant needing some environmental component, such as a chalky soil for beech trees, whereas in fact it simply tolerates it. Tolerance confers a significant ecological advantage because it allows a plant to colonize a site free from more sensitive or demanding species.

Strictly speaking, the environmental need of a species implies a condition required at a specific level to sustain a physiological process. It follows that in the wild the plant will be found only in locations that satisfy such a need. Obviously the same vital requirement holds true when the plant is in cultivation.

In the great gardens of the 19th century every effort was made to provide plants with ideal conditions, while the hoe looked after competitors. On today's development sites we are more likely to utilize the ecological concept of tolerance of adverse conditions exhibited by what have become known as 'tough' species. Modifying a site in one or more ways prior to planting is, however, part of the programme required to achieve many designs (see page 108, Cultivations before and at the time of planting section).

Geographic origins

Today there is a heightened awareness of the possible ecological impact of amenity plantings, not all of it supported by research. As a result of such well-intentioned objectives, those involved may find that they are requested or even required to select plants with reference to one or more of the terms 'population', 'native', 'exotic', 'provenance' and 'accession'.

The spread of a species in the wild is known as its natural distribution. Plants whose natural distribution falls entirely outside any given area are termed 'exotic' or 'alien'. In Britain there are around 2400 'wild' or 'native' species, the majority of which are also found wild in mainland Northern Europe; the remainder grow only in Britain and are said to be 'endemic' to this country. If a site falls within the natural distribution of a species that species is considered native even if it does not grow wild in the local area. For example, in Oxford, seakale (*Crambe maritima*) is technically a native (i.e. a member of the British flora) although it grows wild only on the coast. Species that grow wild in Oxfordshire are termed 'locally native'. The distribution of a species is of course not controlled by national frontiers; however, there is a convention to speak of, for example, a 'British' or 'French' native species.

The term 'provenance' is used to indicate the geographical origin of a population of plants. Occasionally it is used to indicate the conditions under which the specimens were raised. When used by ecologists the term pinpoints the location from which the specimens, or more likely their wild ancestors, originated. A plant species from a particular location may have a distinct gene pool, in which case we may say that the population of the provenance forms an 'ecotype'; see page 41, Ecotypes section.

It is commonly considered among ecologists that on rural sites only native species should be planted, even to the exclusion of their cultivars. The latter ban is particularly pertinent with agricultural crops such as white clover (*Trifolium repens*) or perennial rye grass (*Lolium perenne*), the man-made forms of which differ so greatly from the wild species or 'type' as it is termed. The avoidance of planting some non-native species is linked to two somewhat different concerns. One is the risk that they may produce 'escapes', i.e. feral offspring that become established in the surrounding area, as has Japanese knotweed (*Fallopia japonica*) in parts of Britain, whereas Buddleia (*Buddleja davidii*) has spread in North America, Australia and New Zealand as well as Europe. Pampas grass (*Cortaderia* spp.) has become a pernicious weed in Southern Europe and other warm temperate zones such as California and South Africa. The other concern is that escapes may pollinate the flowers of closely related wild species to produce hybrids and so 'pollute' the native flora.

The term 'accession', applied as a number or code, is used to identify any specific batch of plants or seed of a species collected at one time in one location.

Getting to Know Plants and their Cultivation

While some designers have a deep knowledge of plants, others, though they excel in wider aspects of landscape design, may have only a rudimentary understanding of plants as living organisms and the consequent demands that places on their care.

The plants that bring life to our public, institutional, commercial and industrial landscapes are at their most vulnerable between leaving the production nursery and becoming established within the landscape. The substantial plant science literature upon which much contemporary crop agronomy is based addresses such basic topics as soils, plant-breeding, taxonomy, arboriculture, ecology and plant physiology. This book cites appropriate research findings and recommends their incorporation into 'good practice' where they are transferable to amenity plantings.

Unfortunately there has been little in the way of experiment-based investigation or trials to assess plant behaviour under landscape site conditions. Such as there has been has focused on trees and as a result we have a good understanding of the common causes of poor growth and death among their transplants. Inevitably these studies have identified, brought into focus and blamed commonly encountered site conditions long known by plant scientists and agronomists to be detrimental to plants. For a detailed study of soil–plant interactions, the standard reference remains *Soil Conditions and Plant Growth*, originally from the pen of Sir John Russell (1912) but in its current 11th edition by Gregory and Northcliffe (2013); for an historical perspective, see Thoday (2007) *Two Blades of Grass*.

Anyone wishing to improve their understanding of the challenges faced in establishing amenity plantings will find their time visiting mature landscape schemes well spent. It may be more heart-lifting to visit a great garden, but peruse the local supermarket car park or industrial park and see which plants *really* succeed, which merely survive and which have failed – if you can find their remains! Such sojourns help one to recognize the quality of stock as it arrives on site and to judge its progress at each subsequent stage. These observations may reinforce the argument that under the conditions the opportunity should be taken to use a richer palette.

Great gardens may seem too mature to serve as a guide to new plantings, but they remain one of the very best ways of becoming familiar with a wide range of plants and to see their development as they mature. Many contain refurbished areas in which the date of replanting can be found. Botanic gardens, arboreta and the gardens of the Royal Horticultural Society (RHS) have the added advantage that specimens are labelled; in addition most carry an accession date indicating the age of the specimen. It is wrong to think of our heritage gardens as being full of only old and unobtainable plants; all the best-known gardens have recently introduced new and in many cases 'landscape-useful' plantings.

Nurseries, seed companies and specialist bodies such as the Royal National Rose Society have demonstration areas and/or open days that allow a wide range of plants to be compared and assessed.

Photographic libraries held on the Internet are valuable but their plant portraits tend to emphasize seasonal display and a close-up of flowers, whereas the designer is primary concerned with year-round plant appearance, form and stature. The pictures of pests and disease symptoms, however, are very helpful in enabling users to make an informed diagnosis.

The number of books on gardening is equalled only by those on cookery but few are written for a professional readership, either horticulturists or designers. The further reading which indicates publications that provide an insight into specialist subjects not covered in detail in this book e.g. for sports turf see Christians, *Fundamentals of Turf Grass Management* and Emmons & Rossi, *Turfgrass Science and Management*.

The plant as a component of design

The influence of fashion

Planting design like all other art forms has experienced a succession of fashions. Their analysis and criticism is outside the remit of this book; however, achieving success with plantings within a fashion is central to the author's concerns. As Tony Kendle and Stephen Forbes (1997) point out in *Urban Nature Conservation*, links between initial planting and design style can be seen most clearly by comparing the contrasting approaches adopted when establishing formal and naturalistic schemes.

Formal designs are based on a fixed geometric and quantified order, in which both the exotic and hybrid nature of the plants and their pre-planting nurture support that objective. In naturalistic schemes, the introduced vegetation, either as seed or transplants, is typically of native species and is expected to develop through biological/ecological responses to existing site conditions.

The influence of plant biology

Plant-based designs range from the most natural to the highly contrived, but regardless of site and design the plants retain their fundamental biological functions and requirements. It may not be immediately apparent, but the success of amenity plantings remains indelibly linked to the plant's ancestor's physiology and ecology. There are clear similarities between natural habitats and site conditions. The predecessors of the 'tough' plants that are selected for difficult, environmentally hostile locations are often the stress tolerators, the natural habitats of which are mirrored by conditions on those sites with soils that tend to be nutrient deficient, with limited depth and proneness to drought or waterlogging, such as those brown field locations discussed in the Soil moisture section, page 20).

Combined influences

Four factors influence a plant's establishment in a design and its potential contribution to it:

- The plant's genotype;
- The shape and form of the transplant;
- The husbandry of the plant before and throughout its time on site;
- The site's environmental conditions above and below ground.

Achieving the design intent is a multidisciplinary challenge. It requires the designer to select the kinds and forms of plant material appropriate to both design and site, and the cultivator to select the practice-appropriate husbandry.

The designer should make clear the effect intended when the plantings will reach maturity. There is no one right method of cultivating a plant in an amenity setting; different design features require plants to be treated in very specific ways – for example, when determining the planting density to achieve optimum ground cover or the spacing between trees in an avenue.

A plant's inherited traits must be capable of achieving the size and form required; however, variations in husbandry can produce visually very different outcomes. For example, the plant we know as *Fagus sylvatica* can become a wide, spreading tree of some 60 m in height or, with reduced spacing, a clean-trunked candidate for the sawmill. By exploiting the plant's response to pruning, in particular a loss of apical dominance, we can produce a dense hedge; and further manipulation of both its organs and environment can produce sufficient physiological stress to put it into 'check', resulting in a bonsai-like specimen.

Spacing

The spacing of plants is a prime example of where design and plant physiology interact. To a plant scientist, spacing is all about intraspecific competition. Varying the planting density within a population of a single taxon produces different levels of competition and hence contrasting visual effects.

Spacing-induced competition influences a plant's growth rate, health, longevity and gross morphology. The latter effect changes size and shape through influencing branch number, position and angle to the main axis. These matters are referred to further as they impinge on specific categories of planting.

Associating taxa within a design

Designs must take account of both the amount and the skill of the maintenance the site is likely to receive. It is no good introducing features that will not be successfully husbanded. The aftercare of plantings regarding such matters as pruning, training and thinning is clearly vital if such specific results as coppiced withy willows (*Salix* spp.) and topiary yews (*Taxus baccata*) are to succeed. Providing the client with notes and sketches indicating how the planting is meant to develop may help guide its management – if they ever reach the maintenance crew.

Equally, the siting of both herbaceous and woody taxa make maintenance a consideration in planning for their long-term development. As we will see on page 47 in the Relationship to soil/site conditions section, for the more vigorous competitors interspecific competition is a way of life. In mixed plantings they must be either regularly controlled or allowed to grow and dominate their neighbours. In projects where there is to be only low or unskilled maintenance, a planned evolution to larger specimens of fewer dominant kinds may be the only realistic approach – in which case informal designs are most able to cope.

Standard Contracts and Specifications

The contract document defines the scope and nature of the work required to execute the design. It forms the basis of the agreement between client and contractor.

Plant selection and specification are unequivocally the responsibilities of the designer, and, consequently, the acceptance of the trueness, quality and health of the plants. These tasks are made difficult by the possibility of living material suffering very considerable damage that does not become obvious for some time. Ralph Cobham (1990) in *Amenity Landscape Management* edited the work of six very competent compilers who amassed a vast amount of data on the tasks and processes associated with both the establishment and maintenance of amenity lands. Although the approach is massively prescriptive, it nevertheless provides a clear insight into the considerations that should lie behind the final form of a contract.

Standard forms of contract

Standard forms of contract used in undertaking the establishment of new landscape areas are typically developed from those used in the building industry, as evidenced by Clamp (1995) in the much used *Spons Landscape Contract Handbook*. Most standard specifications are excellent when dealing with quantifiable matters but they require on-the-spot negotiations and often amendments when there are exceptional conditions. For example, the need for irrigation in both quantity and frequency will vary from season to season and year to year. Also, the frequency and methods of weed control required vary depending on soil type and the weed seed bank; neither the latter nor any outbreak of a pest or disease can be foretold – both require on-the-spot decisions.

The importance of selecting the horticultural contractor

The horticultural work arising from professionally designed projects within existing amenity landscapes, such as private gardens and/or estates, and some trusts and public parks, may be undertaken by the client's own staff. In the case of new development sites the horticultural work will usually be done by a contractor – either the main contractor or, preferably and much more commonly, by a subcontracted specialist landscape company. In some instances the client will place an entirely separate contract with a specialist horticultural company.

In all cases the work should be carried out as described in the specification and drawings supplied by the designer. Its supervision may rest with the client's own staff or else the designer will continue her/his involvement through the role of supervising officer. In all such cases the supervising officer's responsibility is to represent the interests of the client by overseeing the quality of the work and its conformity to or permitted modification of the design and contractual documents.

Selection and recommended lists

Horticultural work, from site preparation and planting to aftercare during the defects liability period, should be carried out by horticulturally trained staff, upon whose skill and professionalism the success of the planting will depend. A good contractor is a vital member of the landscape team and should be treated as such.

If the final selection must be based on price it is essential to invite tenders from a shortlist of only companies of known quality. A good contractor provides skilled work in exchange for reasonable profit. It is unprofessional to select a bid that is so low that it cannot possibly provide the skill and care required. Such a mismatch often leads to a state of war between the client's representative and the contractor, in which the biggest casualty is the landscape planting itself.

Vetting and recommending (selecting) the soft-landscape contractor is a critical responsibility and one that, when successful, is of massive benefit to the success of the project. Good rapport and a degree of justifiable trust make for smooth running.

Landscape contractors undertake diverse commissions in which the soft-landscape component may range from detailed gardening to no more than 'muck shifting' on reclamation sites. Do not select on size of company; there are good and bad large and small contractors. Consider small local firms particularly when the requirements are for gardening skills. The contractor's previous experience and skills often relate more to one end of the work spectrum than to the other; commissioning a company with the appropriate area of horticultural expertise, standards and approach to the work is essential. It is important to be assured that the company appointed has a good knowledge of dealing with the kinds of design and range of plants selected, together with experience with the machines, equipment and hand tools likely to be used. Setting aside the obvious need to avoid commissioning incompetent or untrustworthy contractors, there is another important issue. When contractors understand that they are competing for a contract they put in their lowest bid commensurate with a reasonable profit. This is, of course, exactly what a competitive tender is intended to engender; it is absolutely essential, however, that the tender documents make it clear just what standard of work is expected.

A high-standard result requires a high skill level on the part of the staff on site, both those doing the work and those supervising it, and sufficient time and manpower to carry out the tasks correctly and under the appropriate conditions. Contract bids that can be profitable only if they skimp on these essentials bring discredit to the industry and failure to the project; they must be rejected.

Although the subject is outside the remit of this book it would be remiss not to recognize that following the capital works the long-term success of the plantings rests with the horticultural maintenance they receive.

Plant specification

Following the planning authority's acceptance, the proposed plant material is specified in more detail in the schedule of plant material. When completed, that schedule forms part of the contract document issued to the contractor. It specifies the kinds (species and cultivars) of plants required together with their age, quantity, size, form and quality. Whenever feasible, transplants with a form critical to the design should be selected at the nursery. Where such characteristics require any specific 'non-standard' deviations from normal horticultural procedures, such as planting density, staking, pruning and training, the tender documents must spell them out.

Clearly the schedule of plant material is an essential part of the contract document. Whenever possible it is advisable to use a standard format such as those produced by such bodies as the British Standards Institution and the Horticultural Trades Association.

The use of numbers and measurements to define acceptability should be regarded as a first step but should be combined with a judgement of quality. Excellent plants may be a few centimetres smaller or larger than the figure stated, whereas, conversely, plants worthy only of the bonfire may perfectly fit the dimensions. Contemporary horticulture is extremely good at producing 'a crop' – that is, a uniform batch of healthy specimens ready for sale on a specific date. Although the procedures involved are better understood for growing pot chrysanthemums, good hardy-plant nurseries achieve very high standards with the commonly ordered and hence mass-produced taxa. Even the best transplants deteriorate, however, if held in the nursery for too long.

Good uniformity and high quality are often lacking in less commonly ordered plants. Throughout their time in the nursery, such plants have lacked, on the one hand, the batch treatment so beneficial in commercial horticulture and, on the other hand, the tender loving care given to prized rarities in private gardens. Furthermore, because their 'turnover' is so slow specimens tend to be retained well past their prime. In practice, plants are rejected either because they do not meet the specification or an amendment to it that is set out in the purchase order or because they are deemed to be substandard.

Specification and assessment of horticultural work

Horticultural work may be specified and later assessed through the quantification of materials, areas, volumes, distances, depths and even the equipment to be used.

Such a measured approach has its place in determining quality of work: for in-stance, in insuring that tillage or mulch is of the required depth. Craft skill and care are far less easy to quantify, and poor work that can have a great influence on plant establishment and subsequent growth can be hidden as one operation succeeds an-other. One of the objectives of this book is to consider those matters that cannot be easily resolved through standard specifications.

One approach to the specification of horticultural tasks recognizes the reality of the limited time available for on-site supervision. It defines the objectives but leaves the method of achieving them to the contractor within what is usually accepted as 'good professional practice', with the work being assessed at appropriate stages. The advantage of this approach is that it frees the contractor to respond to unforeseen circumstances; the disadvantage, however, is that, by relying on the assessment of 'finished work', substandard work concerning amounts, volumes and depths is either undetectable or else discovered too late for remedial action.

Regardless of the approach taken to site supervision, a central plank of the client representative's professionalism should be the ability to judge the quality of the horticultural work involved in executing the design and in this to be particu-larly aware of pitfalls and problems commonly encountered on landscape sites.

The less the supervising officer's understanding of husbandry, the more he or she is disadvantaged in discussions with the contractor. Typically such discussions are aimed at overcoming any unforeseen difficulties that cannot be resolved by referral to the contract, with its focus on quantification.

Quality of work can easily be equated with neatness and, while of course a clean finish is required, a few minutes with an extra bag of mulch and a yard broom must not be allowed to divert attention from the work that actually influences the growth of the plants.

Landscape Sites

2

Regional and Microclimate Variation

Mean winter temperature may vary within a few miles. This is the most common regional climatic influence on amenity plantings; however, aspect, exposure and topography leading to frost pockets can all add to site-specific environmental contrasts. Because they are likely to influence both design and plant selection, their potential impact should be assessed when visiting the site before the soft-landscape design and plant selection starts. For a detailed insight into a plant's relationship with its immediate surroundings, see Professor H.G. Jones (2014) *Plants and Microclimate.*

Winter temperature

There is considerable variation in both winter average and winter minimum temperatures across regions of the UK. The major influences are latitude, altitude and proximity to the sea. Within these regional variances there are microclimatic variations both between and within sites. Such temperature differences can determine the degree of risk when selecting 'half-hardy' plants discussed on page 48, Hardiness section). Meteorological winter cold data may be expressed as an average condition over a number of winters but damage to plants is a function of the severity of the worst seasons or even of the worst spell of cold during a perhaps otherwise benign winter.

Aspect

Aspect is a significant factor in the performance of plants in gardens in the temperate latitudes and is often thought of in positive terms where a particular location receives above-average amounts of solar radiation. Conversely, there may be areas where the prevailing wind often damages plantings.

In many a garden, skilled plantsmen have used their knowledge to exploit very local microclimates in ways that, in combination with skilled husbandry, allow the cultivation of an unusually wide range of plants. Designers of more-general 'development' landscapes, however, are wisely more cautious about using environmentally sensitive subjects because here aspect often comes into play when it brings adverse conditions including exposure to frost and wind.

Exposure

Most designers of public open space and industrial landscapes prudently select fully hardy subjects relevant to the local climate; nevertheless, plantings on unusually exposed sites can suffer from both early autumn leaf loss and wind pruning, particularly in coastal areas exposed to salt-laden gales. In such extreme locations it is necessary to build up a perimeter windbreak of tolerant species rank by rank, as shown in Fig. 2.1. The outer windward face of such living defences, however, typically presents a truly weather-beaten appearance, as seen, for example, in the surrounds of the Abbey Gardens on the island of Tresco. Such locations pose a unique challenge: their climate is at the same time milder than inland locations in the same area but prone to much stronger winds, possibly accompanied by salt spray. Several books on coastal plantings such as Christine Kelway (1962) *Seaside Gardening* describe how, over time, shelter planting can considerably reduce wind damage, thereby allowing a wide range of plants to be grown. Living shelter is a long-term investment so is not always feasible on landscape schemes as the use of large transplants usually results in failure because they do not establish under such stressful conditions. Where such shelter cannot

Fig. 2.1. Windbreaks. A well planted windbreak both deflects and filters the wind resulting in shelter for approximately ten times its height. Two windward shrubs form the toe of the wedge-like profile.

be used it is advisable to select the most resistant species. Unlike solid structures such as walls and fences, living windbreaks have an irregular profile and are to a degree permeable; both characteristics reduce the risk of turbulence causing damage on their lee side. Temporary non-living windbreaks are briefly discussed on page 61 in the Persistent strong winds section.

Frost pockets

The meteorological phenomenon known as a frost pocket is typically a low-lying area into which cold air drains during radiation frosts. Night temperatures in such places can be several degrees lower than in the surrounding area. This can lead to damage on soft spring growth of trees, shrubs and herbaceous plants and to the flowers of early-flowering species such as *Magnolia*. Frost pockets, such as the one illustrated in Fig. 2.2, can be formed by erecting fences or growing dense hedges across slopes, thereby trapping the flow of cold air.

Phytotoxic atmosphere

Following the increase in urbanization and the intensification of the industrial revolution, many built-up areas suffered from air quality so poor that it proved damaging and even lethal to many plants (let alone people). Indeed, this in part accounts for the prevalence along British city roads and in parks of such tolerant species as the London plane, limes and holly. The passing of the Clean Air Acts all but removed such phytotoxic conditions and has enabled the planting of a far wider range of woody and herbaceous species, as shown by Medhurst (see also page 45, Selecting the taxa – what is a good landscape plant?) in his study of trees in London. It remains prudent, however, to check the conditions on specific sites within industrial complexes – for example, those adjacent to fume vents.

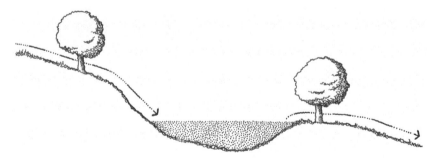

Fig. 2.2. Frost pockets. Cold dense air drains downhill but accumulates in valley bottoms, hollows and behind dense tree and shrub plantings set across the slope.

Inherited Biological Problems

Weeds

To the plant scientist, weeds are agents of interspecific competition. To varying degrees they compete for water, light and nutrients, often with very damaging results. In addition, weeds can destroy the aesthetic intent of a design and indicate a lack of care and managerial interest in the site.

That such competition for water, nutrients and even light can seriously affect all forms of planting from trees and shrubs to herbaceous subjects has been long recognized. Understandably, attention has focused on crops in such overall reviews as L.J. King's (1966) *Weeds of the World*.

Under the editorship of Dr Robert Naylor (2002), *Weed Management* brings together an array of experts that make it a comprehensive study of both the impact and control of weeds. There are frequent changes in regulations controlling the availability and use of pesticides including herbicides. In order to stay within the law anyone specifying or using such materials in Britain should refer to the most recent edition of the UK Pesticide Guide.

Weeds surviving from the site's previous history should be destroyed, typically by using a systemic herbicide, *before* the site is disturbed, or failing that *before* the planting beds are cultivated, a point stressed when considering weed management on page 107, Pre-tillage weed control section.

Pests and diseases

It is of course impossible to eradicate pests and pathogens endemic to the area surrounding the site. Nevertheless, in addition to vetting incoming stock it is advisable to check for pests and diseases that may be already on site, having been carried over from its previous history. Large, vegetated sites may have a significant population of rabbits and/or voles, in which case new plantings must be protected. Retained vegetation may be host to any number of insects and fungi; however, few will transfer to the new stock in sufficient numbers to cause significant damage. Bare-planting sites that appear benign may harbour residual pests such as slugs and vine weevils, which will seize the opportunity to feed on the new source of food provided by the arrival of transplants.

Armillaria mellea (honey fungus) is a common pathogen in woodland and can spread from the roots of infected trees and the stumps of long-dead hosts. It is nearly impossible to eradicate this fungus from standing woodland, but potential infection from isolated victims can be greatly reduced by the removal of the larger roots after felling.

Soils and Soil Fertility

Basic soil science

Soil is perhaps the most confusing – and to some even mysterious – material that may be encountered on site. Soil science is, however, a very well established discipline, with many textbooks addressing the mechanical, physical, biological and chemical aspects of the subject. For a general appreciation of soils and their effect on plants the Gregory and Northcliff revision (2013) of Sir John Russell's *Soil Conditions and Plant Growth* remains outstanding, whereas White's (2005) *The Principles and Practices of Soil Science* gives a clear, concise introduction to the subject.

A fertile soil is made up of minerals, air, water, organic matter and a huge range of microorganisms. Such soils are stratified so that below any surface organic debris is a layer of topsoil, below which is the subsoil. If the soil was formed *in situ*, the subsoil rests on the parent rock or its weathered remains (see the soil profile in Fig. 2.3). The form and nature of the strata are the products of chemical and biological reactions on soil minerals over long periods. Taken together, these layers form the soil profile, the total depth of which can vary from a few centimetres to more than a metre.

The topsoil is of greatest concern to growers as it is typically the most fertile and the darkest layer because it contains most humus. The depth of topsoil

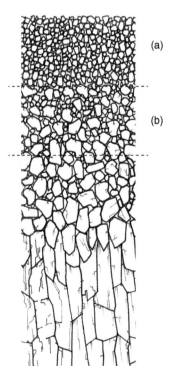

(a)

(b)

Fig. 2.3. Soil profiles. A schematic representation of the profile of an in situ derived soil. In the top soil, (horizon a), sand, silt and clay derived by weathering dominate. In the subsoil, (horizon b), rock fragments are more prevalent.

in lowland Britain is usually between 150 mm and 300 mm; we should note, however, that shallower soils down to a few centimetres successfully support arable crops and semi-natural vegetation types ranging from species-rich downland swards to broadleaf woodlands. Although the greater part of the root system of many plants, both woody and herbaceous, develops within the topsoil, the nature of the subsoil plays an important part in site fertility because it determines water movement and acts as a source of nutrients.

The particles that make up the mineral component of a soil are divided into three grades: the finest forms the clays, the middle grade the silts and the coarsest the sands. Larger particles such as the rock fragments featured in Fig. 2.3 are not regarded as being constituents of a soil; however, they play an important part in determining the nature of the root environment.

The proportions of the three grades, the sands, silts and clays, determine the soil texture and form the basis of soil classification. Whereas the crop agronomist generally deals with *in situ* topsoils and subsoils, landscape sites following ground works may have no such stratification. In such cases concern should be for the nature of the root zone material, be it imported or site derived.

Fig. 2.4 represents the blending of clay, sand and silt to produce a mix most suitable for the majority of plants; these are the loams. A medium loam has approximately 30% clay, 50% sand and 20% silt and not the equal proportions one might think. Loams may vary somewhat, however, in the proportions of their constituents, hence the existence of sandy, silty and clay loams. Each has advantages and disadvantages but all can support a wide range of plants. Samples dominated by either sand, silt or clay are not regarded as topsoils suitable for general crop production but may be encountered on development sites.

The condition, that is, the structure, of a soil at the time of cultivation and planting is far more important than its precise texture, which need be of concern only when one component dominates (for a full explanation see Gregory and Northcliffe, 2013).

Soil structure describes the form and degree to which the mineral particles and humus are aggregated to form peds and the finer crumbs. This structure determines the entry and movement of water and air, both of which are vital to roots. From the cultivator's point of view, a well-structured soil is one that supports vigorous root growth. It has around 70% solid material (including 5% organic matter) and 30% voids containing air and water as shown schematically in Fig. 2.5.

A loam's texture can result in a structure that produces the most desirable solids–water–air ratio; however, a wide range of soils is capable of supporting a wide range of plants.

The primary aim of tillage is the improvement of soil structure; it endeavours to produce a homogeneous, friable surface layer known as tilth (Fig. 2.6). Traditionally, across both agriculture and horticulture, the day-to-day timing of tillage was governed by the weather as reflected by soil moisture. Working saturated land damages its structure, destroys peds and, on heavy soil, produces clods. Furthermore, additives cannot be correctly incorporated when the soil is in this state. In traditional kitchen gardens cultivation so changed the depth and

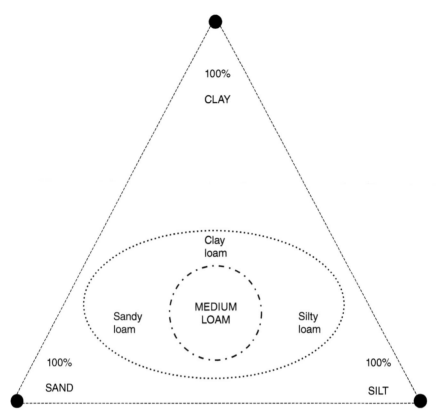

Fig. 2.4. Soil texture triangle. Soil classification is based on the particle size of their minerals. The most desirable, the loams, are made up of various blends of sand, silt and clay.

nature of the topsoil that the entire root system of many vegetables remained exclusively within this zone.

Healthy plants can be grown in very small amounts of soil, as is evident from the success of container-grown plants ranging from poinsettias (*Euphorbia pulcherrima*) to semi-mature trees. These successes are, however, totally dependent on the formulation of the substrate, generally referred to as a potting compost, and a regular supply of water and nutrients. In contrast, long-duration plantings such as orchards, soft-fruit and timber plantations depend on their roots being able to spread through and utilize the whole soil profile. The majority of landscape plantings are of this kind and, given the opportunity, develop the same root systems.

Soil fertility

Soil fertility describes the degree to which the soil is supportive of plant life. Under intensive cultivation, commercial crops are produced in substrates ranging from

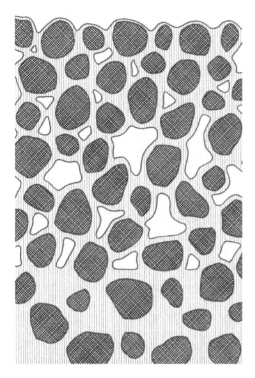

Fig. 2.5. Soil structure. In addition to a skeleton of mineral particles, a good topsoil contains humus, air and water. Minerals and humus aggregate to form crumbs. Water is held within and around these crumbs while above the water table, air occupies the larger spaces.

Fig. 2.6. Soil crumb structure. A tilth of soil crumbs formed by the aggregation of mineral particles and humus forms the ideal planting substrate. Photo P. Thoday.

rockwool to peat and, in hydroponics, water; meanwhile, containerized nursery stock transplants thrive in loamless compost. Diverse as these substrates are, they all supply the plant with its nutrient and water requirements and sufficient oxygen to support root respiration.

Soil fertility cannot be measured in absolute terms; it is relative to the demands of the site's flora. The following are, however, the factors considered by soil scientists when assessing soils for crop production:

- Level of plant nutrients;
- Air content;
- Moisture retention;
- Percentage of organic matter;
- Soil pH.

Plant nutrients

The chemical assessment of soil fertility is expressed through the amount of plant nutrients available. The major nutrients are nitrogen, phosphorus and potassium [NPK]; elements required in smaller quantities are termed 'trace elements' or 'minor nutrients'.

Little has been published on the specific requirements of the many kinds of decorative plants used in landscape schemes and it should be noted that the Department for Environment Food and Rural Affairs (DEFRA) ratings are based on the expected response of arable crops. By comparison, permanent amenity plantings have a modest demand for nutrients.

Air

Soil air makes up around 15% of a well-drained topsoil; its oxygen content is essential for root respiration. Gaseous exchange between atmospheric and soil air is dependent on the pores and fissures within and between the soil crumbs. Anaerobic soil conditions are lethal to plant roots, whether they are caused by denatured mud, compaction or the capping of the soil surface.

Soil moisture

Terrestrial plants, with the exception of those adapted to marshy or boggy habitats, require aerated soil and cannot live in totally waterlogged conditions. Even a seasonally high water table may limit the depth of the root zone. Across the world, millions of hectares of farmland are made productive by land drains. The amount of water a free-draining soil retains is termed its field capacity. In a well-structured loam this is around 15% of its volume. A little over half of this amount, termed the 'plant-available water', is accessible, after which the plant reaches wilting point. Without additional water, plants under these conditions suffer a progressive death of root, leaf and stem tissues. The significance of soil type, the consequences of reduced water availability and the resulting plant stress are all explored by Dr Erich Winter (1974) in *Water, Soil and the Plant*.

Organic matter

The term 'organic matter' covers all materials derived from plants and animals and in all stages of decay from unchanged to humus. Topsoil from meadowland contains approximately 5% by volume. On landscape sites, farmyard manure or more often other sources of organic matter such as composted green waste and leaf mould are used mainly as a soil conditioner. Partly rotted material helps retain both air and water. The humus derived from further decomposition helps to improve soil structure by forming soil crumbs. It is at this stage that nutrients are released and become available to plants.

pH

The pH of a soil is an indication of its acidity or alkalinity. It is measured on a scale from 0 (acid) to 14 (alkaline) with 7 being neutral. Soils typically range between pH 5 and pH 8. Very slightly acid (pH 6.5) soils are considered optimum but the range from 6 to 7.5 is satisfactory for most plants.

Excessively high or low pH is unlikely to have an immediate impact on transplants but can be disastrous in the long run if intolerant species are chosen. Attempts to permanently change soil pH, other than reducing acidity through applying crushed limestone (calcium carbonate), are generally unsatisfactory; it is far better to select compatible species. Many texts list calcifuge (lime-intolerant) and calcicole (lime-loving) species.

Soils on Landscape Sites and their Treatment

With the exception of annual displays, landscape plantings differ from the use of and demands on the soil made by agricultural and horticultural crops. Amenity plantings are often a once in a generation operation, after which the soil is left to develop its natural structure. Under appropriate management such an approach produces excellent conditions for many species.

Clearly the one time when tillage is required is before planting. The recognition that transplanting and reestablishment is a biological challenge to many plants long ago prompted gardeners to provide the best possible soil conditions. On many development sites things are very different; rather than such preparation, soil handling is at the mercy of the overall building timetable come rain or shine.

Soil conditions on landscape sites are liable to differ from those found in nearby agricultural fields and the gardens around old houses, both of which will have been influenced by long periods of cultivation. People used to such situations and the tendency of garden writers to emphasize the specific and elaborate requirements of hard-to-grow plants may well be shocked by the nature and condition of the substrate confronting them, a point stressed by Tony Kendle in the chapter on *The nature of soils on landscape sites and their effect on plants* in Thoday and Wilson (1996) *Landscape Plants*.

Extensive sites may retain areas of undisturbed grassland that retain the local structure, texture and depth of topsoil and subsoil. In such cases they should be cultivated as little as possible prior to planting (see page 118, the section on Planting on undisturbed grassland).

In contrast, previously built-on locations (brownfield sites) often show no recognizable divisions within their 'soil' profile – such as it may be – while many large-scale greenfield developments will have been subjected to massive site disturbance in the early phases of the building works. In practice, most planting locations have both soil structure and profile influenced by groundworks.

Denatured soils

Denatured soils occur as muds or compressed layers. They have lost their structure and they lack the soil crumbs that form tilth. Fig. 2.7 shows a typical area on a development site. It is all too easy to hide such areas under a thin covering of imported topsoil.

Historically the most famous denatured soils must have been those of the battle-fields along the river Somme. Before 1914 these seas of mud were highly productive arable farmland to which, following years of husbandry, they have eventually returned.

Fig. 2.7. Denatured topsoil. An impermeable, structureless layer of mud resulting from denatured soil. As in this case the soil need not have a high clay content. Photo P. Thoday.

On development sites denatured soils usually form a surface layer of mud around 100 or 200 mm deep. This seals in the soil beneath making it anaerobic. The damage is usually caused by heavy vehicles whose passage both destroys soil structure and compacts the strata beneath, making the area waterlogged. Soils can also be damaged by inappropriate forms of cultivation or by their handling and cultivation under inappropriate weather conditions.

Although heavy clay soils are most easily damaged, all but the sandiest are susceptible. Even when removed from a waterlogged site, soils whose structure has been denatured take many months to recover. Building sites are inappropriate places to attempt restoration and tilling saturated soils usually makes matters worse.

Rather than plant into mud it is advisable to remove it, ensure that the exposed material drains freely, and make up any soil deficit. In some cases some improvement to the root zone of less-damaged soils can be gained by incorporating decomposed organic matter to aid water movement and allow roots to ramify between the clods but, as noted in the section on Organic matter, page 21, this approach does not always work.

Bulk density and compaction

Although they have very different appearances, denatured muds and compacted soils both have impeded air and water movement and neither can satisfactorily support root growth.

The bulk density of a soil is its dry weight divided by its volume, usually expressed in grams per cubic centimetre. A bulk density of 1.6 is considered the maximum for satisfactory root growth of many species. Densities above this are impenetrable to roots; the soil is virtually impervious to water and air, and thereby anaerobic. Root-inhibiting levels of compaction are often caused by heavy machinery, particularly in wet weather. Untreated compaction is likely to take many years before it becomes fully eased by natural processes; hence root systems remain confined to cultivated strata or within their planting pits.

Easing compaction

Compacted soils can be eased mechanically but are slow to reform tilth in the form of peds and crumbs.

The term 'ripping' entered the tillage lexicon some 50 years ago to describe the use and action of very large machines fitted with 'winged tines'. These machines are used to break up severe compaction on massive projects, such as the reinstatement of sites following opencast mineral workings. Ripping now describes the work of loosening compacted substrates on any scale. On most landscape sites the larger areas, if free from obstructions and underground services, can be loosened to sufficient depth using an appropriately powered agricultural tractor fitted with a one-, three- or five-tined chisel plough or 'ripper'.

A penetration of 350 mm is desirable, although it may take two or more passes to reach full depth.

Compaction on smaller planting areas can be eased by digging with a backhoe in the hands of a skilled operator. On very small sites it is necessary to resort to using a fork. An alternative to deep tillage to loosen compacted substrata is provided by compressed-air injection accompanied by the addition of some material to keep the fissures open.

Soil depth

Although, as noted above, no specific soil volume or depth can be said to be essential, it is clear that a contractual document requires quantification.

British Standard BS 3882:2007, *Specification for Topsoil and Requirements for Use*, gives a minimum depth of 250 mm for topsoil on areas due to be planted with shrubs and herbaceous subjects. Such a depth may well be based on situations where the root zone is restricted to the depth of topsoil (see page 34, Tree roots section).

A more focused specification will take into account whether or not the planted area is isolated or sits above a root-penetrable subsoil. On many sites, when using the common range of taxa chosen by landscape designers and with the addition of organic matter, the topsoil depth above a root-penetrable subsoil may be reduced to 200 mm. If the design includes trees the overall bed depth may remain at this figure, but planting pits must be dug within the bed to provide greater local depths, as noted in the Planting on undisturbed grassland section page 118).

Both low-grade soil and minimal long-term maintenance indicate the need for a generous root-run because there must be enough root-penetrable substrate to meet the plant's requirements for water and nutrients. Within limits, such a volume may be either shallow but widespread or confined but deep.

Most woody and herbaceous perennials planted into naturally occurring shallow soil or its equivalent on brownfield sites will compensate by developing a wide but shallow root-plate or by exploring the upper horizons of the subsoil.

In confined, isolated beds and planters, with their limited and fixed amounts of soil, both irrigation and fertilizer applications are likely to be needed. When specifying plantings for such containers and isolated planters, it should be remembered that in the long term it is not the number of specimens but the total biomass relevant to soil volume that is the critical factor.

'Roof garden' installations require specialist advice and are not considered in this book.

Drainage

Overall site and hard-surface drainage should have been addressed as an engineering task. Over larger areas surface water can be drained away into the site's substrata via swales or French drains, and the water disposed of through soakaways; this may involve determining the position of the water table.

Waterlogging on soiled areas is closely linked to soil type; the higher the clay content the greater the likelihood of waterlogging. It may be due to any or all of the following: excessive runoff from adjacent hard surfaces; impervious subsurface layers or pans; or compaction owing to heavy machinery (Fig. 2.8).

It is not the water per se that kills roots but the exclusion of air that prevents respiration. The soil profiles of large planted or grassed areas may be drained through flexible perforated pipes set 'to falls' beneath a gravel backfill so as to slope to feed into either a soakaway or a swale.

Ideally the drainage of individual planting areas, typically beds and borders surrounded by the hard surface of paths and roads, will have been linked into a site-wide land drainage system. If no such system has been installed it is impracticable to try to make such linkages during the horticultural works. The impeded drainage of individual beds and borders is usually ameliorated by deep tillage of the whole root zone and where necessary a simple soakaway should be constructed beneath each cultivated area.

Drainage is influenced by both soil texture and structure. Sands are commonly and correctly understood to be free draining, but it is a misconception to regard all clay soils as non-draining. Unless compacted they often have slow but satisfactory natural drainage. To this end, soil structure is most significant

Fig. 2.8. Waterlogged site. Planting on clay loam halted by waterlogging due to surface runoff accumulating where traffic sealed natural drainage fissures. Photo courtesy of Mary Payne.

because drainage takes place via the fissures between the naturally formed blocky peds.

Demolition rubble and toxic soils

Rubble often makes up the bulk of the substrate facing the planter working on brownfield sites. A considerable body of work, much of it emanating from the late Professor Bradshaw's group at the University of Liverpool, has demonstrated that such substrates can support a satisfactory rate of growth of trees, shrubs and herbaceous perennials (see Bradshaw, Hunt and Walmsley (1995) *Trees in the Urban Landscape*; Kendle and Forbes (1997) *Urban Conservation*; and Dutton and Bradshaw (1982) *Land Reclamation in Cities*).

Dutton's fieldwork demonstrated that, for a range of tree species, rather than importing large amounts of soil onto brownfield sites such basic good practices in such areas as planting season, transplant storage and handling, watering and weed control were the essentials to transplant survival, establishment and subsequent good growth increments.

Toxic residues on brownfield sites are mostly associated with industrial activity. Suspect sites must be inspected for traces of materials deemed toxic to humans (see: www.gov.uk/contaminatedland/overview). By law such sites must be made safe before development work commences. It is very unlikely that any permitted traces of such toxic chemicals (substances) that remain will be significantly phytotoxic. Indeed plants exhibit much above mammalian tolerance to both heavy metals and many common organic compounds. Phytotoxic materials on old industrial sites are very unlikely to be regarded as soil and used as a potential root zone. If their retention on site is permitted, they will be buried and capped with an impervious layer before the soil profile is instated above. It should be noted, however, that the surface materials of old industrial sites although free of toxic materials may have a pH above or below the range suitable for most plants, as already noted in the pH section, page 21.

Handling soils

In situ *and imported weeds*

Prior to any ground works, areas due to be stripped and/or planted should be checked for weeds. Perennial weeds including turf and particularly those designated as 'noxious' or 'invasive' should be sprayed out before the soil is disturbed. The vigour of weed growth is a useful indicator of soil fertility and, in rare cases, of toxicity.

To those used to life in long-established gardens, perennial weeds are rarely a problem. In contrast, stored and imported soils may carry perennial weed ramets; the grasses collectively known as couch are by far the most damaging, with bindweed (*Calystegia* and *Convolvulus* spp.) and creeping thistle (*Cirsium* spp.)

not far behind. Perennial weeds can spread underground ('run') beneath the roots of transplants, making their elimination almost impossible. Such weeds spread to ruin the appearance of hundreds of plantings every year and many sites are subsequently written off having been entirely taken over by them. Work schedules may require that such fragments be removed, but their elimination during pre-planting cultivation is virtually impossible and under the combined pressures of cost and contractual deadlines is often not attempted. Attempting to rectify matters later by pulling the tops off the weeds prior to a defects liability inspection is an insult to all concerned.

Soil stripping and storage

Large development sites often engage in soil handling on a civil engineering scale using heavy earthmoving equipment. This makes it possible to design a land form that retains both subsoil and parent material on site, thereby saving on dumping costs. A well-managed site strip will preserve the topsoil for reuse without loss and excessive damage. This requires topsoil and subsoil to be stripped separately. The work must not be done when the soil is saturated because this will lead to loss of structure. The stripped topsoil must be stored free from compaction. Much was made formerly of the need to limit soil heaps to no more than 1 m in height so as to prevent loss of aerobic microflora. There is now clear evidence, however, to show that there is a rapid recovery following spreading from larger heaps, providing that the soil is neither compacted nor denatured when tipped. Topsoil sites should be fenced off and the building trade's habit of throwing washings including toxic substances on to such heaps must be explicitly prohibited. Long development programmes make it likely that weeds will establish on the soil heaps and these should not be allowed to seed; the opportunity should be taken to eradicate perennial weeds such as couch grass.

Treatment of damaged soils

Spreading stored topsoil

When stored topsoil is spread over either subsoil or nondescript fill, no trace of the original 'A' and 'B' horizons remains and there is no continuity with the underlying material.

Spreading usually starts with the soil being loaded by backhoe or digger bucket and transported by dumper truck, two pieces of equipment that are relatively impervious to weather conditions. The soil, however, is not and is more often denatured during spreading than when being stripped. Inappropriate handling may have destroyed structure and reduced soil crumbs to mud or fused them into clods but as the spread soil has unchanged mineral components it retains its original classification (e.g. clay loam, etc.). It is, however, the state of the soil after spreading that determines its value as a growing medium regardless of its location and earlier condition.

Soil additives

Wherever bulk materials are added their effect will be greatly influenced by the physical and chemical nature of the *in situ* soil and the degree to which they are mixed with it. In the case of cloddy or compacted soils incorporation is difficult and slow work. Time must be allowed for this in the contract. As Kendle and Forbes point out in *Urban Conservation* (1997), additives if misused can produce adverse effects. Their use should be determined on a site-by-site basis and directly linked to specific objectives and hence the plants selected.

Organic matter

The need for and advantages of additional organic matter depends on the nature and intent of the design and the condition of the soil. The time-honoured phrase 'well-rotted' indicates organic matter that has lost its original structure but is still far from its final breakdown to humus. This is the stage in which organic matter is most effective in improving aeration and drainage and is also relatively easy to incorporate. For such use the application rate on landscape sites should be increased from the 25 tonnes/ha common in agriculture to 75 tonnes/ha, equivalent to 7.5 kg/m^2 (10 l/m^2) or a 1 cm dressing. In practice quantities of organic matter are difficult to determine and may be best established by reference to the amount delivered to site. Attempts to radically change the composition of a denatured, cloddy and/or excessively heavy soil immediately prior to planting are seldom successful.

Sand

Sand of the right fraction is an effective drainage agent, as is evident in sportsfield drainage schemes. Nevertheless, its use to emolliate clay-textured soils en masse is rarely either feasible or successful because very large quantities are required and must be intimately mixed.

Sand is used in two surface treatments. One is the application of a 100 mm layer to suppress the germination of weed seeds held in the topsoil seed bank. This technique has been successfully used by James Hitchmough when establishing herbaceous plantings.

The other is in the establishment of all-weather sports surfaces. This specialist topic is not covered in this book (see the further reading list) but the techniques involved can be modified to provide durable grass play surfaces in public areas.

Lime

The addition of lime as crushed limestone (calcium carbonate) neutralizes acidity, adds calcium and helps in the formation of a good soil structure by flocculation. Its use on amenity sites is rarely required unless the soil pH is below 5 (see comments in the pH section, page 21). Tests in a soil laboratory can both determine soil pH and calculate the amount of lime required to raise the pH to 6.5. Lime should not be spread on areas in which calcifuges such as *Erica* and *Rhododendron* are to be planted.

Fertilizers – NPK

As a general rule transplant establishment is far more affected by soil structure than by soil fertility and it is not usually necessary to enhance the nutrient level of typical topsoil prior to planting. Landscape schemes do, however, sometimes occur far from the agronomist's fertile fields, and some site substrates are far from 'typical topsoils' and may be deficient even in the minor nutrients.

If, through such indications as poor weed growth, the available material's fertility is doubtful, it is prudent to commission a soil analysis to determine the level of available plant nutrients and the quantities of fertilizer needed to overcome any deficiencies; the required application should be applied during pre-planting cultivation. The laboratory will advise on an appropriate protocol required to collect a truly representative soil sample. Typically such an analysis also indicates the soil pH as noted above.

The most important or 'major' elements required by plants are nitrogen, phosphorous and potassium. We know their optimum amounts, ratios and times of application for the major crops, but no such information is available for the plethora of plants used in amenity plantings. In practice, the use of fertilizers is usually confined to soils or substrates of evident low fertility and/or to recent plantings *where vigorous vegetative growth will hasten the design intent*. In such cases a top-dressing of a general fertilizer, 20:10:10 NPK, at the rate of 20 g/m² at the start of the first and second growing seasons after transplanting, is typical.

As stressed in the Direct seeding of native perennial herbs section, page 127, species-rich meadows and direct-sown exotic herbaceous displays should not receive fertilizer dressings.

Swell gel

The very high quality of today's floral displays in such containers as hanging baskets is often due in part to the use of moisture-holding polymers added to the compost at the rate of one gram per litre. These substances, generally referred to as 'swell gel', slow the loss of soil moisture between irrigation cycles, so preventing the plants experiencing huge swings in water availability. Studies indicate that such polymers are effective when added to the soil used in small raised beds and isolated built-in planting boxes such as those used as street furniture. They may also be applied as a dusting on the roots of bare-root transplants.

Mycorrhiza

Mycorrhizal fungi live in association with the roots of the majority of plants. In nature these near-ubiquitous members of the soil microflora are hugely beneficial or even essential to their hosts through greatly enhancing the uptake of water and nutrients. The use of mycorrhizal cultures introduced as root dips are likely to be of benefit to landscape plantings in two circumstances:

- Where bare-root transplants that have lost most of their roots they benefit because the mycorrhiza both boost the efficiency of the remaining root system and colonize new roots;
- Where planting in previously barren sites, with little or poor soil, because these generally contain an impoverished microflora (R.A. Dutton, personal communication).

Imported (introduced) soils

During civil engineering works, imported soil may have been spread over the site substrate without any attempt to either cultivate it or break down the interface between it and the material beneath it. Any soil-spreading carried out within the landscape contract should be preceded by sufficient cultivation of the existing site material to remove compaction and aid drainage and thereby link the introduced soil into a future planting's root zone.

If the site soil has survived in good condition it may be necessary to introduce only enough soil to make levels up to the height of curbs etc. In such cases it is advisable to mix the local and introduced soils because they may differ in their texture, structure and fertility.

Imported soil may carry a weed seed bank gathered from its place of origin or from badly managed soil storage heaps. As discussed in the Weed control section on page 133, the resultant flush of weeds can become a serious issue during the new planting's establishment phase. As indicated by British Standard BS 3882:2007, *Specification for Topsoil and Requirements for Use,* brought-in soil should be free of perennial weed ramets and should be inspected for such.

The nutritional status of an introduced soil is easy to improve but its pH is not. Failure to match the local soil acidity or alkalinity creates contrasting pockets, which can result in uneven plant performance.

The physical components of 'topsoil' as defined in British Standard BS 3882:2007 are a useful guide when purchasing soil; we must recognize, however, that soils with such a makeup are actually uncommon across most of the British Isles, including large areas of productive farmland, and rare indeed in afforested sites.

Although the most desirable texture maybe that of a loam, wide deviations are perfectly satisfactory, particularly for trees and shrubs, providing the material has a good structure – a point emphasized by Professor Bradshaw and his team at Liverpool University. Texture helps determine structure but other and complex factors are involved. Unfortunately, bulk transportation and spreading inevitably damages a soil's peds and crumbs that are the basis of its macrostructure.

Soil type is certainly one of the determining factors governing plant distribution in nature, even within one meadow. However, husbandry activities such as tillage, irrigation, application of fertilizers, transplant spacing and weed control combine to enable a cultivated plant to succeed in a wide range of soils. It is such husbandry that allows many great gardens to grow the same plants despite being located across a wide range of soils.

Manufactured soil

Reference is made to load-bearing soils as backfill for tree planting pits. When, as at the Eden Project, there was no topsoil on site, more than 800,000 cubic metres of root zone substrate was manufactured. The bulk materials were sized fractions from the mineral spoil from china clay workings, and partly decomposed organic matter. It is not possible to give one formula for soil manufacture because the bulk materials available vary. In general the mineral-to-organic fractions should be in the order of three to one. Such manufactured soils have proved completely satisfactory in supporting a wide range of permanent plantings but, depending on the makeup of the mineral fraction, a balanced fertilizer dressing of 30 g/m^2 may be needed every other year.

The Plants

3

The Conservation of Specimens and Communities

The conservation of plants and plant communities is worthy of consideration on all development sites, whether *in situ* or through translocation. It may be a requirement on sites subject to specific controls such as those applying in Conservation Areas and Areas of Outstanding Natural Beauty. In addition, both greenfield and brownfield sites may harbour protected plant and/or animal species.

Many sites carry planning consent requirements relating to aspects of local ecosystems (habitats) or plant-based features. In some cases features must be retained, in others designs must take regard of the wellbeing of named species. The instructions controlling such conservation work may be issued by external agents such as the officers of Natural England and Historic England and the local tree officer. Outside conservation areas, the regulations concerning the retention of vegetation usually focus on trees and, less often, hedgerows.

In the conservation of woodland, every specimen, regardless of size, of any species with the potential to grow to the stature of a tree is deemed to be a tree. When considering individual Tree Preservation Orders (TPOs) a tree is defined as a specimen with a trunk 75 mm or more in diameter at 1.5 m above ground. Existing trees are considered under three headings:

- Those that may be removed to facilitate development;
- Those covered by a Tree Preservation Order (TPO);
- The remainder of the trees on site.

The protection of trees subject to British TPOs is set out in the National Planning Policy Framework document *Tree Preservation Orders and Trees in Conservation Areas*. The care of the remaining trees and any constraints on site works in their vicinity are usually covered by clauses within the planning consent and are frequently based on the relevant British Standard.

The design team may be required to comply with directives regarding ground works associated with named trees, and clearly these must be adhered to. The extent of the exclusion area around the tree is a case of the bigger the better, having regard to site limitations: a minimum diameter of eight times the diameter of the trunk at 1.5 m above ground is taken as a standard. Such precautions, together with all aspects of the care of trees, are subject to inspection by the local tree officer.

Conservation of habitats

The retention of vegetation may be achieved by leaving undisturbed areas or by reinstating plant material using the translocation techniques noted below. Specific, mainly small-scale areas and their flora, although apparently undisturbed, may be affected by their altered surroundings, for instance by shade from buildings, by drainage being impeded or by ground levels being changed, causing the area to dry out far more than before. In these circumstances it may be necessary to lift the plants, correct the problem and reinstate the plants using translocation procedures.

Tree conservation

The concern for tree conservation and the paucity of research into tree root biology have combined to generate extreme caution to avoid root damage.

The conservation of trees *in situ* must focus on their protection during building works. Notices prohibiting dumping around specimen trees are seldom sufficient protection; exclusion from the contractor's zone and fencing at least 3 m from the trunk is recommended. Where the tree is nearer to construction work, making such exclusion impossible, the trunk should be surrounded with a wrap of chestnut paling or something similar to protect it from damage.

Those responsible for the care and preservation of extant trees are usually very concerned over changes in soil level within the root spread. As already emphasized, the biological impact on the specimen and hence the maximum safe depth of soil to either add or subtract depends on many variables. Most significant are the soil type, the species and the age of the tree.

The issue may be summarized as follows. The spread and depth of a species root system is genetically controlled through plant hormones. In practice root systems are modified by their environment and are found only where conditions of moisture and air are satisfactory. In some soils this may be throughout a considerable depth of substrate; in other cases it will be very shallow, producing a characteristic root plate just below the surface. Removing surface soil or adding an extra layer may move the 'optimum condition zone' up or down. Removing soil over mature roots is relatively harmless, provided sufficient remains to protect them and prevent their drying out. There should always be a minimum of 100 mm of soil above the root zone.

Adding soil becomes progressively more damaging if the material is compacted or structureless, as both conditions inhibit air and water entry. *Loose* fill, including subsoil and 'clean' rubble up to a depth of 250 mm has, in the experience of the author, proved benign on many sites; however, if the trees are under a TPO it is prudent to discuss such changes with the local tree officer.

Where the control of ground works is at the discretion of the designer or the supervising officer it is possible and prudent to apply a more specimen-specific approach that takes into account such matters as the specimen's health and, most critically, age together with the species' root biology.

The covering of a root zone with impervious material is far more problematic – hence the standard recommendation to use a porous grid where hard finish is to go close to the trunks of trees. Trees flourishing in pavements and similar urban hard surfaces illustrate their remarkable ability to tolerate such conditions. Such success is typically based on the ability of the root systems to exploit existing conditions. Covering the root zone of already mature specimens presents a greater risk and requires a very open material. As noted in the section on Planting on sites restricted by hard landscape, page 117, work in Denmark has resulted in the development of a specific soil mix for parking lots. It both provides good conditions for root growth and support for vehicles.

The spilling of phytotoxic substances can result in root death. The base of trees seems to attract washings and it is not unknown for them to be thrown over exclusion barriers!

Tree roots

East Malling Research, formally East Malling Research Station, has a long and distinguished history of investigating tree roots. In the 1960s, Eric Coker excavated a complete apple root system, while in 2014 a group under Professor Peter Gregory carried out the same investigation on an 11-year-old Oak (*Quercus robur*). The specimen had a height of 3.5 m and the limited canopy spread of an oak sapling. It had a trunk diameter of 8 cm and the root spread (root plate diameter) averaged 4 m; a small percentage of the roots reached a depth of 1.8 m. It is reasonable to regard such a root system as typical of an undisturbed specimen growing on a well-structured deep soil.

It has been known for a long time that there is no fixed relationship between the spread of a tree's canopy and that of its roots. The extent and pattern of root spread is influenced by both the species and the situation, most significantly local soil conditions. Professor James Hitchmough (1994), in *Urban Landscape Management*, suggests that under favourable soil conditions the root spread of well-established trees may be at least equal to their height. Ground-penetrating radar is a promising but as yet rarely used technique for the non-intrusive tracing of roots. It is claimed to be able to detect *in situ* roots as small as 2 cm diameter.

The belief that roots are positively geotropic, that is they grow downwards under the influence of gravity, holds good for only a small minority of those that make up the root system of woody plants; the majority are plagiogeotropic. Their horizontal response to gravity produces the shallow root plate characteristic of many woody species. Although in theory the root zone is radially symmetrical, as Professor Gregory stresses roots are opportunistic, and each prospers and spreads outwards or downwards only where conditions of moisture, air and rock density are favourable.

Severe root loss due to ground works can affect the health of the tree and in extreme cases cause death. Nevertheless, each year thousands of street trees recover and grow normally after the excavation of service trenches within their root zone. The transplanting of semi-mature trees has demonstrated that they thrive in spite of the loss of some 80% of their roots when lifted from the field (see page 36, the Translocation of trees and shrubs section). This use of semi-mature transplants has revealed considerable differences in root regeneration between species and the age or more precisely the physiological condition of the specimen also affects such regeneration. It should be recalled that root pruning was a standard practice in fruit tree management to help control vigour and improve harvest.

Root systems of woody plants consist of permanent and temporary roots. Permanent roots increase in length and diameter over time and provide the basis of a root system. It is from their more distal regions that the thin 'fibrous' roots are produced; these do not thicken but are the main gatherers of water before being sloughed off. The severity of the effect of mechanical damage to roots depends on the tree's ability to regenerate new roots. Following root loss, roots are produced from the proximal regions of the root system by cambium cells located both along the length of the permanent roots and at the ends of severed roots.

The 2 cm limit of detection by ground penetrating radar may be of significance in some cases. While the smaller roots at the distal extremes of the root system will be lost when roots of this size are severed they are among the most prolific regenerators. By contrast to such recovery, if soil within the root zone is so denatured and/or compacted as to cause root death, no recovery will occur.

Stability is most affected when the root loss is all on one side of the tree. For any given distance from the trunk, the potential loss of roots is far greater per length of cut when a trench runs at a tangent to the tree rather than directly towards the trunk.

Service runs wherever possible should be threaded between or around trees; however, where they must pass close to retained specimens they should be hand-dug to minimize damage because it is often the case that pipes and cables run below the root plate. The old technique of back-filling the area around severed roots with friable soil remains good practice. Horizontal boring equipment capable of passing service cables or pipes beneath trees is available; however, its hire cost limits its use to very special locations.

Translocation of trees and shrubs

Prior to the last third of the 20th century, recommendations for tree translocation emphasized preparing the specimen for one or two years before the move. This involved severing the roots at a distance of 1–2 m from the trunk and back-filling the resultant trench with 'good' soil to encourage new roots. At the time of moving, a trench of slightly greater diameter was hand-dug and the rootball secured and lifted onto a low-loader to transport it to a previously prepared planting pit. This approach was successful but involved a great deal of work. Today, moving a semi-mature tree from location to location within the same site is more likely to be done using heavy machinery such as a 'tree spade'. This equipment undercuts, lifts, transports and repositions accessible specimens all in one operation. In such cases the root plate is typically reduced to less than 2 m in diameter, resulting in more than 80% of the roots, measured by length, being lost. As noted above, such successful practices contrast with the prohibition on digging even archaeological trial pits anywhere within the supposed root spread of a protected tree.

There is some disagreement over pruning at the time of translocating large plants. One view is that canopy thinning reduces the risk of loss of water from transpiration; others argue that the plant needs all its foliage for photosynthesis. It is likely that each treatment has merit depending on circumstances. Transplants with ample roots and irrigation will benefit most from leaf [stem] retention whereas those with no irrigation and few roots benefit from pruning.

Specialist advice should be sought on the feasibility and cost of relocating young healthy trees that fall foul of development works when if moved would contribute to the proposed design.

Hedges

Mature hedges have been successfully translocated metre by metre carried in an excavator's huge front-end bucket. In some cases, hedges and individual specimens are coppiced before moving with the intention of stimulating subsequent growth as illustrated in Fig. 3.1. With moves of only a few metres, good results have been achieved by excavating a way to the new location and using a bulldozer to push the hedge onto its new alignment. Hedge translocation has the benefit of moving the complete hedge habitat and its flora and fauna; this is surely the main justification for translocation rather than establishing a new hedge from whips.

Translocation of species-rich grassland

The herbage is first cut low before the sod is undercut, lifted, transported and placed as turves on the prepared surface of its new location. Where the

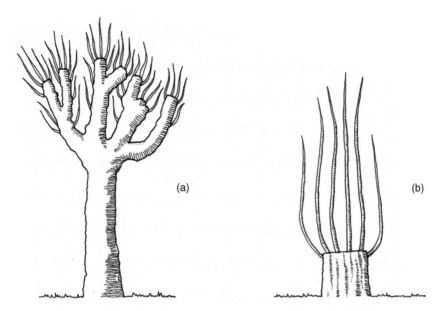

Fig. 3.1. Pollards (a) and coppice (b). Trees and shrubs may be either coppiced or pollarded before translocation. Many species will regenerate from the cut surfaces if moved with sufficient roots.

sods do not form a continuous cover the spaces should be blinded over with the appropriate topsoil. Both forb-rich grassland and heathland comprising mainly ling and heather have been successfully translocated as sod. At the Eden Project when translocating lowland heath generous amounts of soil were taken along with plants of *Erica vagans* ensuring that part of the original site's seed bank was also moved; subsequently many seedlings added to the success of the undertaking (Fig. 3.2).

Advanced planting

Typically this term is used to describe the establishment of structure planting several years before a site is developed. The planting is usually in the form of tree screens and shelter belts that it is hoped will provide some enclosure and/or identity and sense of place to the development. The technique can be applied with equal success to establishing semi-natural grassland, making it more resistant to wear and tear immediately following development. Advanced planting offers the advantage of being able to use forestry techniques free of the constraints of a development site and of the need to achieve a neat appearance and instant impact.

Fig. 3.2. Habitat translocation. The translocation of *Erica vagans* lowland heath by sod and seed at the Eden Project. Photo courtesy of Tony Kendle.

Background to the Plants We Use

4

Introduction to Taxonomy and Nomenclature

Following on from the work of Linnaeus, Rae and others in the 18th and 19th centuries, plants and animals are classified into ever more closely related groups. The groupings we are concerned with are the family and within that the genus (plural: genera) – containing in its turn the species. Advances in molecular biology and genetics have allowed us to make this system more and more closely linked to what is considered to be each taxon's evolutionary pathway. This system of progressively more closely related groups is of considerable help in both designing and maintaining plantings because closely related plants tend to require similar conditions and respond similarly to husbandry inputs.

Nomenclature

Any system of classification requires its subjects to have names, and the giving and use of such names to have rules. Those rules are the prerogative of nomenclaturists, and for naturally occuring species, be they in the wild or in cultivation, they are set out in the *International Code of Botanical Nomenclature*. The *Code* covers the taxonomic rules governing ranks above and including that of species. An important section deals with the grammar used in the writing of scientific names – the family, genus and species. The two rules of particular relevance to our interests are that family and generic names are *always* spelt with an initial capital letter, whereas specific names (epithets) are *always* spelt entirely in lower case.

To find information on the naming of variants found within a cultivated species or its hybrids we must turn to another document, the *International Code of Nomenclature of Cultivated Plants*.

Taxonomy

Many of the plants we use on amenity sites arrived into our huge exotic flora in the form of species collected from their native habitats; others were already being cultivated, in which case they may well have been forms selected for specific qualities. Subsequent cultivation, selection and cross-breeding has added many more forms, with the result that the flora available to designers has the full range of the divisions into which taxonomists group plants.

Species and their Derivatives

Species

A species (note that this word is both singular and plural) describes the specimens of any community of wild plants that have a similar appearance and genetic makeup.

Within the population there will be slight variations. When such variations are quite clear and breed true the group in which they occur may be identified as a subspecies. Botanically, differences not significant enough to warrant the designation subspecies are termed varieties and are important in decorative horticulture, both in their own right and because they often serve as the basis for cultivars; in some cases they may be described using the non-technical term 'good form'. The majority of the trees used in larger-scale and rural plantings are species, and up to the mid-20th century this was also the case with shrubs. Today the selected forms, the cultivars of both trees and shrubs discussed below, are more likely to be chosen for urban plantings. Few wild species are chosen from among the herbaceous perennials, with the exception of some grasses; however, we meet exotic species in Professor Hitchmough's designs as is evident in Fig. 4.1, a photograph of one of his stunning plantings in London's Olympic Park. Only native species are used in the mixes sown by Landlife in producing 'wildflower' floral meadows.

A further reference is made below to the role of species in specific locations when discussing the terms 'native' and 'exotic' with regard to plants.

Subspecific variants

Without going into the agonies experienced by some taxonomists over the definition of a species, it is evident that many species are made up of several clearly distinguishable forms or subspecies perhaps evolving towards a fuller separation from the type. Some subspecies are particularly suitable for specific landscape

Fig. 4.1. A fresh approach to planting in public greenspace. One of James Hitchmough's highly successful plantings in the Olympic Park, London 2012, based on a succession of summer flowering taxa. Photo courtesy of James Hitchmough.

plantings, for instance *Gleditsia triacanthos* f. *inermis* as a street tree and *Juniperus rigida* subsp. *conferta* as ground cover.

Ecotypes

Typically a species with a wide natural distribution will have developed 'local' discrete variants, each limited to a specific provenance. Such variants are termed 'ecotypes'. Their genetic variation is not sufficient for them to be considered as subspecies and they will almost certainly breed with the 'type'. It therefore follows that any introduction of either the type or a foreign ecotype will endanger future generations of the local ecotype.

If it is deemed important to maintain the unique nature of the ecotype it alone should be used locally. There is a point where many people would regard this concern as esoteric. We now know that it is usual for many communities, even if isolated by just a few kilometres, to have some genetic variation from neighbouring communities, though many of such are not differentiable in appearance from the phenotypes. For example, among grasses we find heavy-metal tolerance on mine spoil heaps and wear tolerance on footpaths.

Clearly the nursery industry cannot supply such variants 'off the peg', but if it is wished to perpetuate the gene pool, as perhaps on an Site of Special Scientific Interest

(SSSI) following some site works, a programme of seed collection or even lifting, nursery storage and replanting might be attempted, as noted on page 32, the section on the Conservation of specimens and communities.

Varieties and forma

The term 'variety' describes wild plants showing a minor deviation from the typical appearance of the species. 'Forma' is a term less commonly met; it describes those forms of a species that show the least but consistent variation from the typical.

Cultivars

Taxonomically speaking, the rank of cultivar is restricted to cultivated forms genetically distinct from their precursor species known as the 'type'. Cultivars are often promoted for a specific feature such as overall shape and form or the quality or colour of their flowers or fruit. Such single developments may not, however, be the only changes that the genotype has undergone, making it necessary to become familiar with mature specimens so as to be certain that they will serve the design. Some remarkable single-feature improvements such as flower size have appeared on abysmally poor growing types; such forms may have a place in some gardens but not in the tough conditions of the many landscape schemes where the robustness of the whole plant is paramount. Cultivars of trees, shrubs and herbaceous perennials are usually propagated by vegetative means, creating a clone and so avoiding the need to genetically 'fix' them. The situation is not, however, quite as clear-cut as that because some perennial plants are also raised from seed. Seed-raised populations usually show some slight variation that, while not mattering in informal designs, are a problem where strict uniformity is required. For a deeper insight into how the development of cultivars has played an ever more important role in our use and appreciation of plants see Thoday (2013) *Cultivar: The Story of Man-made Plants*.

The origin of cultivars

Cultivars come about in several distinct ways. While the vast majority arise as a result of either mutation or hybridization, they may also originate as a result of any one of the following:

- A vegetative (somatic) mutation – a 'sport';
- A germline mutation resulting in fertile seed;
- A hybrid or 'cross' between two genetically distinct individuals;
- A chimera – a form of mutation producing an individual with cells of more than one genetic origin;

- A polyploid – an individual with more than one set of chromosomes;
- A virus infection resulting in a change in an individual's appearance.

Mutations

A mutation is a change in the makeup of a gene or genes. It may either occur naturally or be induced. Mutations result in changes in the appearance and/or the function of the specimen or its offspring.

Somatic or vegetative cell mutations occur during the division of cells in the growing point (the meristem) of a bud. Such mutant shoots are often referred to by horticulturists as 'sports'; their offspring form a clone.

Germline mutations occur during the cell division that produces the germ cells or gametes. Germline mutations occur naturally but can be induced. Most resultant seedlings are sterile so to form a new cultivar an individual must be cloned by vegetative means.

Chimeras

A chimera, as any Greek scholar knows, is a monstrous beast whose body is made up of the parts of different creatures. In the plant world, many chimeras originate with a somatic mutation but continue to have cells of both mutated and original forms functioning alongside each other within the same tissue; this produces such effects as variegated and distorted foliage.

A few chimeras get very near to realizing the myth: they result from the tissues of two plants fusing while their cells remain unchanged. Grafting can produce such oddities, hence the coining of the misleading term 'graft hybrid'.

Hybrids

Most people associate hybridization with the crossing of two species, but the term hybrid describes the offspring of a mating between any two genetically distinct individuals of any taxonomic rank: cultivar × cultivar; cultivar × species; species × species. Hybrids may occur naturally, but the majority result from deliberate breeding programmes. Such 'crosses', as they are generally termed, can be selected and named as either true-breeding or vegetatively propagated cultivars.

F1 cultivars

As Mendel showed, hybrid seedlings from crosses between pure-breeding parents are uniform in their first daughter or filial (F1) generation. This makes it possible to take advantage of a combination of good qualities bestowed by the parents. As subsequent generations will not come true, F1 cultivars must either be propagated vegetatively or be remade for each sowing.

Ploidy

Ploidy describes the number of sets of chromosomes in each cell. The typical condition is termed 'diploid' (one set from each parent). Those with extra sets are termed 'polyploid'; the increase in the sets of chromosomes often results in greater size, vigour and yield. Many wild species are polyploids.

Infections and monstrosities

Attractive as they may be, some cultivars are simply diseased. The ornamental Japanese honeysuckle, *Lonicera japonica* 'Aureoreticulata', is a well-known example, while the striped tulips that were at the heart of tulipomania are surely the most famous of these infected plants.

Synonyms

Synonymy implies that a plant is known under more than one name with all the confusion that that can cause. Synonymy, or more precisely its rectification, is one of the main causes of plant species having their name changed. This is most irritating when a well-known name has to be superseded by an obscure but older one to conform to the International Code of Botanical Nomenclature. Among ornamental plants the enthusiasm for new introductions led perhaps inevitably to exactly the same cultivar being traded under several names. Such confusion, accidental or deliberate, goes back more than 200 years. Sometimes it happened when a plant of unknown pedigree was named (or renamed) by a new owner but there has been a lot of straightforward plagiarism both by gardeners and by nursery firms.

There is what one may call the reciprocal of synonymy, where several forms of a plant, some better than others, are offered under the same cultivar name. This can be is extremely difficult to detect in young transplants and may result in a reduced performance later. This is yet another reason why it is important that stock is sourced from only the most reliable and knowledgeable nurseries.

Selecting, Assessing and Purchasing Landscape Plants

5

Selecting the Taxa – What is a Good Landscape Plant?

Britain is blessed with a climate that allows us to grow a huge range of plants. Nurturing the demand for this cornucopia are tens of thousands of skilful private gardeners and the professional staff of the National Trust, Historic England, private trusts and some public authorities. Success with many of these plants requires skilled inputs and the creation and maintenance of specific and very localized conditions.

From this rich assemblage of flora a much-reduced number have proved themselves to be appropriate for landscape planting; they must be able to both thrive and satisfy the design intent with limited attention. These species or their selected cultivars, available in significant numbers and appropriate sizes, are listed in various publications including the HTA's (2002) *National Plant Specification* (NPS).

Several wholesale nurseries produce lists of plants suitable for specific locations and uses, as does the very authoritative *Hillier Manual of Trees and Shrubs* (2014) together with clear descriptions of the plants. Also the *Darthuizer Vademecum* (2005) published by the Dutch nursery Darthuizer Boomkwekerijen is particularly good at selecting the 'tough' plants needed on some landscape projects. Some lists bring together plants that share a biological requirement or, more often, tolerance; for example, there are lists of lime-tolerant species but the plants included may have nothing else in common and may require very different growing conditions and serve completely different design functions.

Books containing lists of particular value in landscape projects include Keith Rushforth's *The Hillier Book of Tree Planting & Management* (1987), *The Dynamic Landscape* edited by Nigel Dunnett and James Hitchmough (2004) and Richard Hansen and Friedrich Stahl *Perennials and their Garden Habitats* (1993).

Beginners should take heart that it is less damning to be criticized for using the same plants over and over again than for being responsible for disasters emanating from selecting inappropriate species. As plantswoman Mary Payne says of many 'garden treasures', 'They may be beautiful but they have no will to live' – and, I would add, certainly not on a development site. Nevertheless there is a time and place for the more unusual. That unique annual publication edited by Janet Cubey, *The RHS Plant Finder* (2015), lists some 75,000 taxa that are available within the nursery trade in Britain in either wholesale or retail numbers and prices, and also those plants that have received the RHS Award of Garden Merit (2015): www.rhs.org.uk/plants/trials-awards. This award takes some pertinent issues into account but again be on your guard – a supermarket car park is not a garden!

In ecological terms, to grow successfully a plant must be adapted to prosper under the conditions prevailing in the spot in which it has been planted. All plants are 'tough' in their *natural* environment; otherwise they would have become extinct. It is not the aim of this book to dissuade people from trying to break away from the handful of taxa that have become the near ubiquitous choice for public, institutional and commercial plantings; there are under-used winners out there, as John Medhurst's study of London's street trees available on his website has revealed: www.johnmedhurst.landscape.co.uk/trees.

Fashion massively influences plant selection. On large-scale rural projects such as new roads we find that, while in 1930s the Roads Beautifying Association were enthusing on the planting of exotic flowering trees and shrubs, by the 1990s The Department of Transport's *Good Roads Guide* (1992) recommended only native species in the countryside. On smaller projects, the second half of the 20th century saw a marked bias towards evergreens (the 'winter never comes' approach) rather than planting deciduous taxa to more clearly celebrate the change of seasons through both foliage and flowers.

'Landscape plants' must support few pests and pathogens. They should be fully hardy for the location and tolerant of local weather conditions: for example, the high, salt-laden winds of coastal sites. They must deliver the aesthetic requirement of the design in almost every year and not just, for example, in atypically warm, dry summers.

Although a specific requirement such as 'ground cover' may be uppermost in the designer's mind, it will be only one of any plant's attributes. The possible selection may also require a condition not found on site or may need to tolerate one that is prevalent. For example, the genera *Pachysandra*, *Hedera* and *Vinca* contain many popular ground cover plants. Most are woodland species, however, and while we may select them for their ground-covering growth habit their physiological requirement of at least partial shade must also be satisfied if they are to contribute fully to the design, a consideration stressed by Professor James Hitchmough in his introduction to that very informative compendium Hitchmough and Fieldhouse (2004) *Plant User Handbook*.

Relationship to Soil/site Conditions

On the basis of plant responses to their surroundings, including their interaction with other plant species, ecologists have identified contrasting survival strategies that produce plants with distinct growth forms and habits. These ideas are explored by Professor Philip Grime (1979) in *Plant Strategies and Vegetation Processes.*

'Stress-toleraters', which include many semi-desert and mountain species, are slow growing but able to 'hang on in there' in spite of limited resources. As a result they can be very useful in difficult sites. The garden literature is rich in comments on tolerance to such environmental conditions as salt winds and lime-rich soils and these are valid and helpful when dealing with specific locations. Garden writers, however, very rarely encounter the kind of conditions met with on many development sites and so have little need to focus on tough, stress-tolerant kinds of plants.

'Competitors' are lush, fast-growing species; these, as the term suggests, typically outgrow other species but they are prone to suffer badly under adverse conditions. Fast growth is a quality that is both admired and feared in garden plants – praised through such descriptions as 'a good doer' and dreaded like 'a thug that chokes its neighbours'. In nature, vigorous species (the competitors) can make great use of good growing conditions, ample light, water and nutrients. If the site can support such demands, competitors return the complement by providing a rich display of vegetation; it is often wise, however, to use them in isolation from other low-growing herbaceous and woody species. Such plants are miserable and unattractive on stressful sites.

Spread is not necessarily linked to great vigour, being based on the production of ground-level or below-ground stems or roots that generate small offsets. Some spreaders make excellent ground cover but others go bare in the centre as the original plant dies out. Spreading species can become invasive when they overgrow their neighbours; hence the most adventurous should be confined by hard landscape or surrounded by taller-growing species beneath which they will be shaded out.

The opportunistic ruderals, typically annuals or short-lived perennials, prosper in competition and stress-free sites but when adverse conditions come they die, leaving behind their seed. Many colourful 'bedding' plants are at heart ruderal and the more extreme their adaptation to this condition the shorter their lifecycle and hence their flowering period. The bedding plants with the longest flowering seasons are in fact perennials.

While such responses have evolved to be advantageous in nature, they may or may not be beneficial to the plant under the conditions and in the positions in which it is placed on site. A plant's morphology, anatomy and physiology are all genetically programmed to attempt to develop under the dictates of this inherited strategy. A mismatch between a plant's genetically ordered growth style and its surroundings inevitably leads to its failure to live up to expectations as a part of the design.

Growth rates, vigour and spread

Planning authorities and clients both frequently ask for an estimation of how long a planting will take to deliver a specific design intent or what it will look like in a given number of years.

Clearly there are great differences in the growth rate between species of woody plants. The fastest-growing deciduous tree species growing in the British lowlands come from the genera *Alnus, Betula, Populus* and *Salix,* all of which when young may increase in height by 1 m per annum. The growth rate of woody plants and hence their annual increase in height varies with their age, with the fastest growth of many trees being when they are between 5 and 12 years old. Valuable data on tree growth rates appear in the Forestry Commission's publication on the potential of 44 native and non-native tree species by Willoughby and others (2007).

Local site conditions can have an enormous effect on growth rate. These 'limiting factors', as they are termed, include both conditions beyond the control of site management such as exposure and rainfall, and those that can be influenced such as weed competition and planting density. Such environmental conditions can reduce woody plant growth to a few millimetres per year by sending transplants into check as research by the Forestry Commission showed in *Trees and Weeds* by Davies (1987).

The annual horizontal spread of some low-growing 'ground cover' species can be equal to the height gained by upright forms. In other cases, spread is based on specific structures such as rhizomes or runners. It is these growth forms that can become invasive by growing into the space allocated for other species.

Stress tolerance

Hardiness

The term 'hardy' is loosely used to describe a taxon's tolerance to the range of environmental conditions discussed above as encountered on the site in question. It more often specifically addresses a plant's tolerance to cold. Such a tolerance is of course linked with the conditions under which the plant or its ancestors grew in the wild.

Many of our decorative taxa originate from milder climates and hence some are classed as 'tender' (requiring a glasshouse). In contrast those from regions with winter temperatures consistently below freezing are termed 'fully hardy' (can withstand sub-zero conditions). The intermediate and ill-defined group are known as 'half-hardy'. In reality the world's flora presents a continuum of cold tolerance.

As pointed out on page 12, the Winter temperature section, the cool-temperate climate of the British Isles exhibits a wide range of winter temperatures, not only at the regional and urban heat islands but even at the microclimatic scale. Hence a 'half-hardy' taxon may flourish in one location and be killed in another. Although many private gardeners like to 'risk' dubiously hardy plants and are prepared to provide winter protection, such a policy is unlikely to be appropriate on landscape sites. Nevertheless one should not abandon all consideration of the many plants classified in the general literature as 'half-hardy'. Even within the small compass of the British Isles, experience shows that some species, while being extremely risky in some areas, are completely dependable in others. Hardiness is therefore not simply regional but even site-specific. It is advisable to supplement the literature's vague definition of half-hardy, based as it must be on a blanket consideration of 'average conditions', by turning to local knowledge and observation.

Several systems that divide land masses into hardiness zones have been developed. The most frequently used is that introduced by that great American botanist Alfred Rehder in his *Manual of Cultivated Trees and Shrubs Hardy in North America* (1940). Several equivalents have been developed for Europe including Britain; these are available via an Internet search on 'Plant hardiness zones in UK'. Few would disagree that these maps are useful but rough guides and, as Rehder points out, they cannot take local let alone microclimatic conditions into account. Clearly the climatic maps are no use without a link to the plants. The hardness guide used by The Royal Horticultural Society is the best but typically errs on the side of caution, with the result that many plants can be grown in what gardeners like to call 'favourable' sites in zones one division colder than indicated on the map. As suggested above, however, such optimism is probably best avoided on most development site landscapes.

Shade tolerance

In many gardening books, comments on 'shade plants' (sciophytes) are confined to the extreme conditions where only plants such as ferns will prosper. On development sites shade is most likely to be from buildings, which give a form of partial shade where direct sunshine is excluded but without an overhead canopy of foliage. In addition to shade tolerance, such curtilage locations may require plants ranging from climbers and wall shrubs to herbaceous species with year-round quality and where possible a seasonal display. Among suitable perennials are bergenias and hellebores, whereas annuals include various begonias and *Impatiens*.

Drought tolerance

The most familiar method of plant survival under desert conditions is succulence, as found in the water-storing cactus; but as the song reminds us, the sage brush grows with the cactus. It is to the same class of mostly evergreen, leathery or hairy-leafed xerophytes that the Australian eucalyptus, the Mediterranean cistus and lavender belong. Such resilient dry-period toleraters are valuable where there is limited soil or where shelter from adjacent buildings reduces rainfall.

Waterlogged soils

Only specifically adapted plants can avoid root asphyxiation when water drives out the air from soil. Such conditions may be found on landscape sites either because of failure to drain localized areas or because areas have been designed deliberately to experience intermittent periods of saturated soil or partial immersion. These conditions, found in natural wetlands, are of increasing importance within larger developments as water-balancing zones or else as catchments for surface runoff. In a garden setting such 'bog gardens' may be carefully manicured but on landscape sites it is usual to treat them as semi-natural habitats – either as water meadows cut over two or three times a year during dry periods or as species-rich wetlands with such tall-growing forbs as meadowsweet (*Filipendula ulmaria*) and yellow loosestrife (*Lysimachia punctata*), while coppiced coloured-bark willow provides a woody plant option. Unintended wet places within more formally designed areas may be due to ground water or impeded drainage following compaction or denaturing of the soil, a problem discussed more fully in the Denatured soils section on page 22.

Weather tolerance

Any landscape plant selected in part for its seasonal display should have a good chance of performing each year. In a garden it is part of the genre to gamble on the display organ of perfectly hardy subjects 'being lucky'. We hope that spring frosts will spare the *Magnolia stellata*, summer storms the delphiniums, rain the zonal pelargoniums and autumn gales the *Acer palmatum* leaves, but in public areas it is prudent to select only the most dependably 'weatherproof' subjects.

Lifespans and durability of design

The concept of the individuals of a species having an average lifespan is generally recognized. Under cultivation an individual's length of life may be extended beyond that typical of its species. The greatest human-induced longevity must surely result from coppicing, with one lime (*Tilia cordata*) in the arboretum at Westonbirt, estimated to be c. 2000 years old.

Alongside their biological lifespan, cultivated plants may be thought of as having either commercial-crop or aesthetic lifespans. Our interest is with the aesthetic, that is, the length of time a plant will remain attractive before it deteriorates visually.

Landscape subjects should have a long aesthetic lifespan; few public open spaces or commercial sites are refurbished in less than 10 years.

Low maintenance plantings

Landscape plants must require little husbandry in either time or skill to deliver the design intent. Woody types should not require regular skilled pruning to produce the desired shape, form or texture and, most importantly, size. The need for permanent support rules out all but self-clinging climbers in most landscape schemes. In this respect many herbaceous garden favourites now have shorter or 'sturdier' free-standing cultivars worthy of selection.

The ability to thrive for many seasons without replanting is virtually a requirement when selecting herbaceous perennials. For example, cultivars of *Geranium*, *Hemerocallis* and *Anemone* × *hybrida* have an undisturbed life *in situ* of at least 10 years without deteriorating, their annual maintenance being little more than a late-autumn cutting down.

Aesthetic and safety considerations

Detailed consideration of a plant's contribution to the aesthetics of a design is not the aim of this book but there are some considerations that are more pertinent in a public landscape than in a private garden where care and attention can overcome shortcomings. The selected plants must contribute to the design all the year round or be invisible when not doing so. Many well-loved and admired garden plants such as roses have one season as the beauty and one or more as the beast. Another unfortunate tendency is for the display of some otherwise excellent plants to 'die ugly' – that is, for their flower heads to remain attached, requiring dead-heading if they are not to distract from later blooms. In this respect, cultivars of *Camellia* × *williamsii* are preferable to those of *Camellia japonica*.

Poisonous plants

Both our native and our exotic garden floras contain many poisonous plants. Although we have lived with them for centuries, it is wise to avoid their use in public locations.

The RHS has compiled a list of poisonous plants cultivated in Britain. Thanks to British gardeners' love of exotic species, the list should be of use in many countries. It should be noted that this list includes many commonly grown species such as yew (*Taxus baccata*) and cherry laurel (*Prunus laurocerasus*) that cause illness only when specific organs are ingested in large quantities; hence those who specify the plant, may wish to make their own judgement as to their suitability. In addition to toxic plants, those plants that have sap or pollen that can cause distress should be avoided in places where people congregate, for example, children's play grounds and old people's homes.

Seasonal display

Where seasonal display is important to the design, both its duration and the appearance of the plant or plants before and after its moment of glory should be considered.

Avoid plants that need dead-heading to either ensure more flowers or because the dead flower petals do not drop as in the case of some camellias noted above. The flowers and autumn leaf colour of most trees and shrubs are ten-day wonders. Berried fruits last from two weeks to six months depending both on their makeup and their attraction to birds. The length of the flowering period in herbaceous plants, both annuals and perennials, varies from a few days to several months, and while the short but sensational types may produce a wonderful annual event in the garden, something more sustained is required in many public locations. The seasonal display of any one plant in the garden is usually part of a sequence of such events which together give a long period of interest. In public landscapes, single 'mass' events such as the flowering of Japanese cherries are employed and one of the most frequently encountered are mass *Narcissus* plantings to serve as harbingers of spring.

Growth rates and mature size

Woody plants may become both out of scale and/or too large for the site long before they reach the end of their aesthetic lifespan. The inspirational designs created at shows and garden festivals must look their best within a short time of planting. They often depend on both close planting and the appearance, beauty and elegance of immature specimens, for example young cedar trees, which if allowed to mature would completely change if not destroy the design. At the other extreme are the fabled tree plantings of the 18th-century landscape designers who planted for maximum effect 100 years thence.

Plant selections that grow fast and deliver the design intent for a year or two then become a problem are inexcusable. However, the *ultimate* size of a taxon may not be considered critical when it is intended that the planting will be refurbished before size becomes a problem. This may seem a very unprofessional approach but it is defensible in locations where the development is identified as having a short lifespan. On such sites perennial plants, both woody and herbaceous, need not be regarded as immortal, but may be selected to be successful for a recognized period, say 10 or 20 years.

In some cases size may rightly be expected to be controlled by pruning, as with hedges or coloured-stemmed *Cornus*. Many out-of-scale plantings are due, however, to the wrong choice of taxa, for example, broadly spreading shrubs that nevertheless reach 2 m in height being used as ground cover, resulting in sight lines being obscured and services and footpaths obstructed. Such bad selection invites the very unfortunate twice-a-year overall clipping of what was intended to be a natural-form planting.

Selecting and Assessing Cultivars

It is generally understood that the large differences between species within most genera are critical when selecting plants for a specific location and function but choice of a suitable cultivar from within a species may be as important. Selecting those that have proved to be successful on earlier projects makes sense, but *only* if conditions are sufficiently similar to warrant such a copycat approach.

A very high percentage of the shrubs and herbaceous perennials used in amenity plantings are cultivars drawn from the far greater number introduced for and cultivated in our private gardens. Cultivars differ in a range of characteristics other than the obvious, such as flower colour, for which they may have been selected. These differences may include important landscape qualities such as growth habit, length of flowering period, form, texture and, as in the case of cultivars of *Aucuba japonica* and *Skimmia japonica*, whole-plant vigour. Disease susceptibility can have a profound influence on the success of a planting – for example the use of one of the scab (*Venturia inaequalis*)-resistant *Pyracantha* Saphyr series in an area prone to this disease.

In many cases the true species has been eclipsed by one of its 'improved' cultivars – as for example with *Viburnum tinus* 'Eve Price'. Therefore when the true species or indeed a specific ecotype or provenance is required it must be clearly stated in the plant schedule (for an explanation of these terms see the Geographic origins section on page 4).

Unlike in the realm of the world's major crop plants, there has been very little investigation into the biology that lies behind differences between cultivars of decorative plant species. Nevertheless, however vague, such comments as 'a strong grower' may be worthy of note. Some cultivars were introduced if not specifically for their value in landscape plantings nevertheless with such sites very much in mind, as in the case of Edward Scanlon's 'Tailored Trees' and some of the cultivars selected by the Dutch company Darthuizer Boomkwekerijen.

The selection or avoidance of variegated cultivars deserves a special comment. Some have sufficiently strong 'other-than-green' foliage to contribute a specific colour to areas of the design. Many plants, however, that have individual leaves that fully warrant their close-up photograph in the catalogue, when seen from a distance become a mass of dull and fussy foliage. Many variegated cultivars that are derived from chimeras revert and in the absence of the appropriate maintenance the mutant character for which they were selected is lost (for an explanation of chimeras, see page 43). Plants whose spring foliage is a uniform yellow are often listed under variegated forms. Their leaves may provide a pleasing seasonal highlight for several weeks before turning green.

Aids to selection

Photographs can be valuable aids to cultivar selection, but are often in the form of 'plant portraits' or even close-ups of the display organ at its most showy. Shots of whole, mature, unpruned specimens or planted groups give a better indication of the plant's value in the landscape. In this respect wholesale nursery catalogues tend to be of greater help than many gardening books. Assessments such as those that lead to the conferring of an Award of Garden Merit by the RHS are not made lightly but are focused on the plant's performance and value in a well-maintained garden (www.rhs.org.uk/plants/trials-awards). A small number of reference books are particularly helpful when selecting cultivars for landscape plantings. For trees and shrubs, see Wandell (1989) *Handbook of Landscape Tree Cultivars* and, for herbaceous subjects, the Joint Liaison Committee on Plant Supplies (JCLI) (1981) *Herbaceous Plants Exotic and British Native.*

Selecting and Assessing Individual Stock Quality

Stock quality is considered in British Standards 3936 *Nursery Stock Specification* 2013 and guides such as the HTA publication *National Plant Specification* (2002) or its more recent iteration and proformas for contractual documents (see: www.csdhub.com) (2015). These guides indicate the usual categories into which stock is graded and/ or offered for sale. Such a numerical approach as, for example, by container volume is extremely useful in providing a sound basis for stating what is required and for rejecting undersized material but is less use as a guide to either plant quality or health. It is easy to overemphasize the importance of the gross morphology of transplants as a definitive guide either to their successful and rapid establishment or to their ability to develop into the form of the taxon envisaged in the design intent. For example, there is no discernible difference in appearance, after two growing seasons, between one planting of whips of the same age and health, initially graded to a length of 700 mm, and another of 1000 mm, if both batches receive the same treatment.

Selection by size of transplant

When specifying tree transplants the most commonly used character is the trunk circumference measured in centimetres at 1 m above ground level. However, it should be noted that trunk girth cannot be directly linked to transplant height as the relationship varies between species and the methods of production. There may be an enthusiasm to achieve as mature a design as possible immediately following planting. To this end the nursery stock industry offer a wide range of plants at a larger than typical transplant size.

Several factors should be considered when specifying transplant size. Although larger transplants make a significantly greater immediate impact, they still cannot reflect the scale of the design at maturity.

The selection of large transplants in the desire to produce an instant effect may not maintain this immediate advantage; the larger and more starved of water and nutrients a container-grown transplant has been the greater the risk of it going into check for one or more growing seasons (see the Starved or 'in check' stock section, page 62). Bare-root or balled-and-burlapped (rootballed) large specimens that have lost a significant percentage of their roots may respond in a similar way if they are not kept well watered.

Larger shrub and herbaceous perennial transplants tend not to spread 'into' one another and so fail to form a continuous canopy across the group.

Any significant increase in transplant size will have a considerable impact on the budget and it may be necessary to explain the cost benefit or lack of it to the client.

The rate of growth of young shrubs varies greatly and hence the time taken to reach full or, in design terms, meaningful size. Hence a 1 litre *Cistus* x *hybridus* or *Hebe* 'Autumn Glory' can reach full size and deliver the design intent in 2 years, whereas some skimmias will take three or four times as long. It is commonly found that smaller tree transplants of around 1 m in height catch up and overtake larger specimens such as 3 m 'light standards'.

The temptation to plant at a high density to produce the appearance of complete cover typically results in overcrowding, which within a short time produces a patchy and uneven appearance discussed on page 7, the Spacing section.

The care needed during the defects liability period is different for large and small transplants. Small transplants are at greater risk both of weed competition and of damage and disturbance through careless weeding. Large transplants often require more frequent irrigation during the first two growing seasons. Poor or insufficient irrigation can result in a tree carrying less foliage and a thinning of the canopy. As a result, such a specimen makes less impact in its first years on site than it did in the nursery.

Specific considerations when assessing stock

Seasonal influence on transplants

Quality assessment is not a beauty contest and over-emphasis on the perfectly shaped transplant may detract from less obvious but important factors that can contribute to its establishment. Amenity transplants are now traded and planted all year round and as a result they may arrive on site in any stage of their annual growth cycle. Hence the appearance of such organs as leaves, stems, buds and roots will vary, as will their vulnerability to environmental conditions such as drought, cold and wind. As a general guide, plants in active growth in late spring and early summer are most at risk.

The roots

Roots are attacked by a wide range of both fungi and invertebrates (see the Pests and diseases section below); the surface of the rootballs of container stock should be checked for dead, damaged or infected roots.

ROOT DEFICIENCY. Plants have a species-specific optimum root–shoot ratio. Loss of root stimulates the plant to attempt to replace it and so restore the balance. From the evidence of excavated total root systems noted on page 34, Tree roots section, it is clear that both bare root and burlapped tree transplants when delivered to site carry but a small proportion of their original root systems. Dormant-season-lifted specimens not only survive, but, with the correct care, will make around 300 mm of extension growth in the growing season immediately following. Root regeneration is discussed on page 34, in the Tree roots section.

Specimens lifted from the field and containerized should not be offered for sale until their root systems have re-established. Under good nursery conditions, remarkably little root is required for containerization to be successful, but such plants are not satisfactory transplants until the species' typical root–shoot ratio is on its way to being re-established; this applies especially if the plant is in leaf.

The value of fibrous roots on such transplants, although much vaunted by traditional gardeners, is to some extent problematic and varies from one kind of plant to another. Such roots appear to be important in nurturing conifer transplants but of course are useless if dried out and dead. In contrast, they play only a minor part in the establishment of those species that readily generate new root from older thicker roots, as do most of the Rosaceae – a point shown many years ago by the American horticulturist H.M. Stringfellow.

EXCESS ROOT. Left to their own devices, plants adjust their amount of living root by controlling their root –shoot ratio; at the same time the plant controls both the radial spread and depth of the developing roots. It is the unnatural procedure of container growing that prevents this control and can produce a pot-bound root system with roots spiralling round the inner face of a container. This condition is shown in Fig. 5.1. The air pot in Fig. 5.2 prevents such a root mass from developing; the tip of each protuberance is open to the air causing protruding root tips to dry up and new laterals to develop behind them. The pot-bound condition can become a problem after transplanting in any one of the following ways:

- The mass of spiralling thin roots will not grow out to colonize the surrounding soil and are extremely vulnerable to dehydration, while in the short term they are the most important gathers of water for the transplant.
- Root system extension will come from the thicker permanent roots within the soil ball but if any have reached the edge they can become coiled-spring-like

Fig. 5.1. Pot bound transplants. After a year many transplants will have produced masess of root around the inner face of the container. Such specimens may show symptoms of nutrient deficiency and reduced growth. Photo P. Thoday.

Fig. 5.2. Air pot. The open tips of the protuberances on the ridged plastic cylinder encourage radial root spread rather than forming a spiral. Exposure results in their drying off but new root forms behind them. Photo courtesy of Andrew McIndoe.

structures that in time may result in poor stability, much increasing the risk of the plant blowing over in strong winds. To overcome this risk it is common in some countries to make several vertical slits down the sides of the rootball at the time of planting thereby severing any spiralling roots. It is claimed that transplant establishment is not adversely affected.

- Pot-bound specimens are simple to detect if the taxon produces roots of a normal thickness, but those with hair-fine roots such as *Erica* and *Rhododendron* are easy to overlook. Techniques used to overcome root spiralling and being potbound are noted on page 74, in the 'Stock prepared for sale in large containers' section.
- It is difficult to wet a compacted rootball. The densely packed rootball produced by container-grown stock is the transplant's only source of moisture. Failure to keep it moist, even with irrigation, is a common cause of plant death, because once having dried out such rootballs are very difficult to rewet in the open ground. This problem is most often found in stock transplanted from large containers.

Foliage

As a general guide, foliage of a transplant should be typical of the taxon in size and colour; however, there are some additional considerations. The foliage of good-quality juvenile specimens will tend to be larger and the leaf margins less irregular than that of mature flowering specimens with which the inspector may be more familiar. This juvenile foliage should not be confused with the excessive soft growth noted below, page 62. Plants lose by far the greatest part of their water through their leaves so it follows that the more foliage a transplant has the faster its rootball will dry out.

Stems

Deciduous shrub transplants should have between three and six sturdy, not etiolated, current-season's stems with an internode length typical for the juvenile growth of the taxon. In many species these stems, or 'breaks' as they are known, should go on to become the plant's permanent framework. Fewer than three suggests a weak plant, whereas more than six will probably have developed as a result of cutting back; none may be sturdy enough to form the basis of a permanent framework and indeed many deciduous species do not develop from the stems you buy at all but from buds at the base of the plant (see page 83, 'Transplant forms available' section. The stems of deciduous subjects should be of mature [ripe] wood with well formed lateral and terminal buds.

Flowers

It was observed long ago that once some plants come into flower the amount of new shoot or root growth decreases and, if transplanted in this condition, they are slow to re-establish and grow away. It is now recognized that many plants, both woody and herbaceous, have such distinct vegetative and reproductive phases. Hence when herbaceous perennials are transplanted in the growing season after flower buds have been initiated they rarely grow away vigorously for the rest of that year.

Annuals are usually transplanted with their first flowers showing colour. This is fine for most types because they will still make the essential vegetative growth and spread needed to fill the space allocated; however, flowering should be no more advanced than this. Rather than poor growth, some annuals, particularly if they have been overcrowded in their seedling stage, respond by producing a premature apical inflorescence followed by rapid senescence and death.

Pests and diseases on bought-in stock

Crop protection pathologists regard a pest as any organism whose presence causes a significant reduction in crop performance. Clearly this view does not extend to include every fungus, bacterium and invertebrate animal that may live or feed on the crop plant. This damage assessment approach holds good in amenity plantings where the amateur gardener's concern over the slightest blemish is irrelevant and indeed counter to the contemporary interest in biodiversity.

The remedial measures associated with the long-term maintenance of planting are not considered here. Landscapers are directly concerned with those pests and diseases that arrive with the transplants and those that attack the plants very shortly after planting. The aim should be to prevent the introduction of serious troubles through careful inspection, not least of the roots, because once infected stock is planted pests can do a lot of damage before they are detected. When inspecting stock beware of specimens that have had stems cut out for other than routine nursery pruning. Pathogens often spread from a first point of infection before symptoms become evident.

Plants should be rejected if on arrival on site there are symptoms of the pests and diseases noted below. Further details of these pests and diseases, the means of their control and the common symptoms by which they may be recognized, are available from the Royal Horticultural Society via their website.

Root aphids

Several species of aphids live on the roots of container-grown plants. The insects are easily seen and recognized by their white mealy coating when the rootballs are removed from the container.

Vine weevil (Otiorhynchus spp.)

The damage caused by adults of species of this genus is easily recognized as notches bitten from the edges of the leaves of a wide range of herbaceous and woody plants. This damage disfigures the plant's appearance but it is the eating of the plant's roots by the larvae that either cripples or kills the plant.

Following any evidence of the appropriate signs of either leaf or root damage or the discovery of the adult or larval stages of the insect, the batch should be rejected. Should the damage become apparent after planting, the soil around the roots of the affected plants should be treated with an approved insecticide.

Slugs

Slugs and snails are, on occasion, introduced from the imported stock but there may also be a residual population on site. These may cause a great deal of damage to newly planted stock, particularly herbaceous ground cover such as hostas.

If the problem is considered likely to be limited to the establishment phase, apply slug control pellets on top of the mulch when damage becomes apparent. Where conditions suggest that the site will carry a large slug population it is better not to use highly susceptible plants.

Rose diseases

Roses suffer from several damaging diseases of which the two most likely to be found on transplants are rose mildew, causing a white downy cover on young leaves and shoots, and black spot, which results in dark patches on foliage that subsequently falls resulting in near leafless shoots. Infected plants should be rejected. Outbreaks during the defects liability period should be controlled by spraying with an approved fungicide.

Box blight

The symptoms of this disease result from attacks by either or both of two fungi. It has become more prevalent on box (*Buxus* spp.) in recent years. It causes the death of both individual branches and whole plants. It is first recognized by the bronzing of patches of foliage followed by leaf fall. As with *Pythophthora* (see below), pruning off infected stems is ineffectual. Infected plants should be destroyed and the consignment rejected.

Phytophthora spp.

Species of the genus *Phytophthora* cause perhaps the worst of the common fungal diseases of shrubs. They attack a very wide range of species of all sizes, including young transplants, typically killing the plant stem by stem. It is tempting for both nurseries and site gardeners to cut out the infected shoots so as to leave an apparently healthy specimen; in practice, however, this is rarely the end of the problem because the fungus will have already spread within the plant. Specimens detected on arrival or during the defects liability period should be replaced.

Fireblight (Erwinia amylovora)

Happily the worst of the doom-laden predictions of the disease caused by this pathogen have not materialized, but it remains a real problem among such members of the Rosaceae as *Cotoneaster* and *Pyracantha*. Stock usually shows infection by the death of individual shoots; later the pathogen spreads throughout the plant. Needless to say, infected stock should not be offered for sale but it happens, in which case the nursery will most likely have pruned off the infected shoots but almost certainly not have eliminated the pathogen. Beware of pruned susceptible taxa. If this disease is confirmed either before or after planting the consignment should be rejected and replaced. Infected plants should be removed and burnt.

Introduced weeds

Although this may seem a trivial matter, it can become the source of a very serious and image-damaging problem. Perhaps the biggest offender of those spread by seed is hairy bitter cress (*Cardamine hirsuta*); it is almost always associated with container-grown stock that was allowed to become weed-infested. Though cleaned up before sale, the compost often carries seed that germinates on site. The plant then spreads rapidly by successive seedings. If infected stock is detected at planting time it should be rejected; otherwise the only solution is to remove the top 10 mm of compost. Perennial weeds including rhizomatous grasses can be carried in the rootballs of field-grown stock. Such infections can become the nucleus of bed-wide infestations that are extremely hard to eradicate and, as mentioned on page 30, Imported soils section, it is prudent to reject such material.

Damage due to environmental conditions

Sun

The trunks of young trees, that is to say the cambium cells within them, can be damaged by overheating when in direct sun. Recent transplants are most at risk as there is a reduced sap flow within their vascular system. Sun scald is very unusual in the UK but of real concern in Southern Europe. It is said to be worse when the transplants have been shaded by their neighbours in the nursery. The usual symptoms are sunken areas on the stem with the cambium brown and dead beneath. The recommended precaution is to wrap the trunks in some insulating material before planting.

Extreme cold

Container-grown, less-hardy taxa may have their root systems killed while standing outdoors in the nursery in a severe winter. As the tops may be unaffected, the damage may not be apparent without removing specimens from the containers and inspecting the intact rootballs. Temperatures of −30°C can damage the cambium cells within the stems of many species that are otherwise regarded as hardy. Such temperatures can even cause the bark to split. Such damage is rare in the UK and hence plantings are not usually wrapped for winter protection as they are in Central and Eastern Europe. Imported stock may be already so treated.

Persistent strong wind

Wind pruning typically affects mature planting. Clearly it is most dramatic when tree branches are wrenched off and leaves shredded and removed in a storm. Persistent wind from the same quarter can shape the overall canopy of a tree or shrub by damaging buds and/or very young shoots, resulting in a lopsided specimen.

With recent transplants, the limited or confined root system reduces water uptake and makes them particularly vulnerable to foliage and bud desiccation. When there are persistent winds such water loss can lead to the drying out of the whole plant. This is a widespread problem, particularly in the first part of the growing season in a dry year on unirrigated sites.

Temporary windbreaks, typically of plastic net or strip, designed to remain in place for around 2 years, are a great asset in establishing plantings on very exposed sites; regrettably, they are an ugly intrusion into an otherwise pristine new layout.

Waterlogging

Root systems can be killed by waterlogged conditions of only around 1 week's duration. This can happen during the dispatch and pre-planting storage of stock.

As with frost damage, such root death may not be apparent without removing specimens from containers and inspecting the intact rootballs.

Physiological disorders

The processes that control the health and growth of plants are based on the supply of raw materials (water, minerals and carbon dioxide) and a favourable environment providing sufficient light, warmth and soil air. Failure to provide for any *one* of these will result in the plant failing to flourish or dying.

Although the quantity and balance of these requirements vary between species there is a basic level that most species need to survive and establish in a new location.

Starved or 'in check' stock

While each environmental factor mentioned above may cause specific damage to recent transplants, physiological stress, often termed 'check', may result from more than one factor and can be exacerbated by loss of roots and the holding of stock in inappropriately small containers or in a nutrient-poor substrate. It is particularly evident and of greatest concern in tree transplants, where a reduction in essential biochemical processes results in the production of small leaves and the near cessation of the growth of shoots and roots. This state of suspended animation may persist for several years, during which time the plant slowly dies; even if it recovers it might still fail to achieve the design intent.

Soft growth

At first glance the foliage and shoots of plants that have been grown 'soft' – that is, in low light and possibly high levels of nutrients, humidity and heat – have the same appearance as those in a juvenile state; this can lead to confusion. Soft growth is very susceptible to damage by cold, drought and desiccating winds. Soft stock should be hardened off (acclimated) before leaving the nursery. The hardening-off process may result in some reddening of foliage and slight leaf scorch; neither need be cause for concern.

Purchasing Stock

It takes many months, or in a lot of cases years, to produce saleable plants. Nursery stock growers propagate the kinds and numbers they think will be in demand, but there is always a risk that stock of a specific kind will not be available in the number and size called for in a design. Forward ordering, by which a nursery is informed of a pending order, is an excellent way of reducing last-minute problems. Such arrangements may be drawn up as a formal contract but, because of the uncertain nature and timing of many developments, forward orders may be nothing more than statements of intent. Binding agreements in the form of contract growing put the intended purchase onto a firmer footing but transfers the risk of wasted stock to the client.

Nursery visits give the opportunity to select and reserve specific batches of plants or to tag individual specimens for the most critical locations. In most cases designers or their horticultural contractors will deal with wholesale nurseries but many designs call for the planting of small numbers or even individual specimens of unusual plants. To obtain these 'one-offs' it is often necessary to turn to one of the many small retail nurseries listed in the *RHS Plant Finder*. When faced with a complex planting scheme it is usually cost effective to use the services of a plant factor, often a nursery owner, who offers the service of acting as a 'middleman' and assembling the required plants into a single order. This person then becomes the supplier responsible for the quality and condition of the plants that pass through his/her hands.

Substitutions

According to the International Code of Nomenclature for Cultivated Plants, cultivars must be unique and have been 'selected for a particular attribute or combination of attributes which may be morphological, physiological, cytological or chemical'. Without doubt there are situations where such 'particular attributes' are deemed essential to the design but this is not always the case.

It is not unusual for a specific plant to be unavailable from the chosen supplier. It is best for the client or the client's representative to state in the contract documents how such a matter is to be resolved. In cases where the client intends to purchase plants directly from a nursery, any substitutions must be agreed in writing before the plants leave the nursery. When a landscape contractor is responsible for obtaining plants he must seek and gain the permission of the client for any substitutions before incurring any cost to the client.

When a named plant is out of stock most suppliers offer an alternative cultivar and, when there is only a *slight* difference in, say, colour of flower or height of transplant, and if the alternative brings the same qualities of growth habit, vigour and ultimate size to the design, then the substitution is usually agreed.

Incorrectly named stock

Stock can get muddled or mixed in the nursery and in transit. Labelling is vital; it is impossible to distinguish between many cultivars of the same species of dormant bulbs, other herbaceous subjects and even shrubs. Transplants muddled before planting lead to mixed groups and a loss of the design intent. Mix-ups should be sorted as soon as they become apparent.

Wrongly labelled stock, where an unwanted taxon carries the name of the ordered plant, should lead to rejection on delivery if the error is clearly identifiable at that time. Evidence of incorrectly supplied plants that shows up during the first growing season – that is, within the defects liability period – presents the supervising officer with the right to request replacements. Replacement typically sets back the design intent by a year, so if the difference between the requested and the supplied taxon is slight it may be preferable to overlook the error.

Travel, import and plant variety rights certificates

There are three reasons why stock may carry some form of certification:

1. It may be on the register of endangered species. Regulations enforced by most countries require imported specimens to conform to the requirements of CITES (the Convention on International Trade in Endangered Species of Wild Fauna and Flora). Although often associated with exotic orchids and cactus, CITES also applies to several hardy species including some palms and the tree fern *Dicksonia antarctica*. In such cases individual specimens must have the required documentation.

2. It may be an import requiring a phytosanitation certification. The international trade in plants clearly runs the risk of transferring pests and diseases from one country to another. An international system of plant passports linked to phytosanitation inspections aims to ensure that plants are free from specific pathogens.

3. The plant may be a cultivar registered and protected by plant variety rights. The cost of such specimens includes a small levy payable by the nursery to the breeder on each specimen sold.

The responsibility to conform to these regulations and to have provided the required documents rests with the plant supplier.

Understanding Nursery Practice

6

Most of the things we buy are produced within the closed world of a factory and it is of little concern to us to know if the object has been made by a person or a robot. Nursery stock is markedly different.

The nursery stock production section of professional horticulture has played a long and esteemed role in plantsmanship. Nursery owners have been responsible for the introduction of many species and their cultivars. In the absence of interest from research establishments, nursery men and women have contributed much to our understanding of the cultural requirements of many decorative taxa. They have long been regarded as partners in the quest to understand and use plants in the garden and wider landscape. Figures 6.1, 6.2 and 6.3 show stock ranging from extra heavy standard trees to container grown shrubs.

For generations designers have visited nurseries to be inspired and informed and to select stock while working on planting plans ranging from Chelsea Flower Show gardens to new town roadscapes.

The terms used in the nursery can indicate how transplants are produced and that can influence the state and form of the specimens at the time of sale.

Amenity plants, ranging from bedding to forest trees, are produced by wholesale growers operating within the fully international nursery stock industry. In their production programme, plants may pass through several owners and not infrequently through several countries.

Fig. 6.1. Large-scale shrub production. Open-air standing ground with windbreak. Container spacing and individual irrigation contribute to extremely uniform high-quality plants. Photo P. Thoday.

Fig. 6.2. Stock production under polytunnels. Protection gives container-grown young stock excellent growing conditions. Specifically designed plastic structures with roll-up sides provide maximum ventilation and prevent over-heating. Photo courtesy of Andrew McIndoe.

Fig. 6.3. Field production of trees. These are well-grown extra heavy stand-ards prior to lifting; note the weed-free site and the wide spacing allowing the development of well-formed crowns. Photo courtesy of Mary Payne.

Propagation

Methods of propagation and their long-term consequences

Seed

Seed is used in nurseries to propagate true-breeding taxa. In practice, this includes several very different groups of stock. These range from trees and some, mostly native, shrub species to British native and some exotic herbaceous perennials such as hellebores to the multitude of true-breeding cultivars of bedding plants.

Seedlings differ from vegetatively raised specimens in several ways. The root systems of very young seedlings, particularly those of woody plants, tend to have a taproot with smaller lateral roots growing from it. In most cases the dominance of this taproot is soon lost; however, in order to achieve a more branched root system by the time they are sold, they may be undercut during their nursery cultivation.

The seedlings of most species go through a juvenile phase in which they grow vigorously but do not flower, a stage that in woody species typically lasts for several years. Clearly this is an advantage in establishing a tree and shrub screen or hedge but it can slow the production of seasonal displays of flowers and fruits.

Vegetative or asexual propagation

Vegetative or asexual propagation is an option when raising many species but is essential for those cultivars that are not genetically fixed. With the possible exception of grafting, whose consequences are noted below, the method selected by the nursery has little if any effect on plant performance after planting.

Vegetative propagation derived originally from a single specimen creates a number of genetically identical individuals – a clone. In practice such propagation is achieved by cuttings, layers, division or graftage.

CUTTINGS. Most of the kinds of stock used in landscape plantings will have been propagated by cuttings. With few exceptions the cuttings will have been made from fragments of stem taken from the 'mother' or stock plant. Propagation by cuttings takes several forms and is carried out at various times of the year in locations ranging from heated glasshouses to outdoor beds. On average, cuttings take two to three years to make a saleable transplant. At first, cuttings develop a different root system from seedlings. They tend to have several equally dominant roots, rather than the seedling's taproot that develops from the primary radical. Such differences are lost during the first year or two and play no significant part in the performance of the transplant.

LAYERS. In this form of vegetative propagation the stem of the propagule remains attached to the parent plant when the new roots are produced. Although the method was of real significance in the past it is now seldom used in commercial nurseries. Transplants derived from layers should behave just like those produced from cuttings.

DIVISION. Historically, division entailed the breaking up of clump-forming stock plants to produce large transplants. Plants derived from a single ramet develop in a similar way to cuttings. In the main, division has been superseded by the use of cuttings in most large-scale commercial nurseries.

GRAFTAGE. Grafted plants are produced by joining the stems of two genetically distinct but related taxa; the lower part is known as the rootstock, the upper as the scion, from which develops the required specimen. Figure 6.4 shows two of the most common forms of grafting woody species. The scion on the whip graft is typically around 100 mm long, while in chip budding it is reduced to a single bud. Following the 'carpentry', stock and scion are bound together until united. Grafting is widely practised across horticulture; however, from the landscape point of view it is generally restricted to woody plants that are either difficult to multiply by other means or else slow to make saleable plants.

(a)

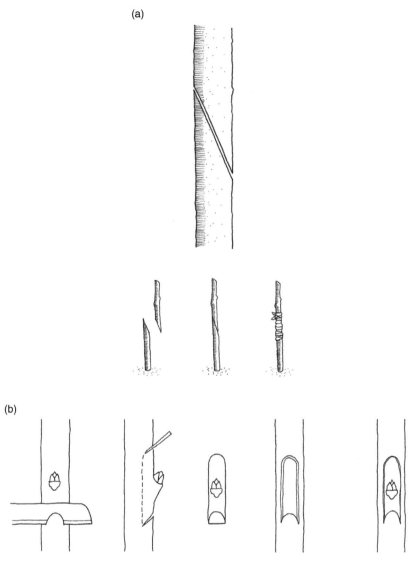

(b)

Fig. 6.4. Grafted specimens. A graft consists of a scion joined at the union to a rootstock. All subsequent stem growth comes from the scion while the stock develops into the root system. Whip grafting (a) and chip budding (b) are two common methods used to propagate trees and shrubs.

Propagation by grafting may have several significant long-term consequences. The rootstock influences the vigour of the scion. The rootstocks of decorative plants are commonly seedlings selected in part to increase the vigour of the resultant plant.

The worst potential drawback with grafted stock is the tendency to produce shoots from the rootstock. Known as suckers, they can obliterate the scion. They typically arise where the rootstock is markedly more vigorous than the scion or where there is partial incompatibility. The standard work on plant propagation is Hartman and Kester (2002) *Plant Propagation: Principles and Practices*. For a detailed review of grafting, see Garner (1947) *The Grafter's Handbook*.

Post-propagation Nursery Husbandry

As already noted on page 56, Excess root section, container-grown stock held for too long at the nursery may become pot-bound and/or nutrient-deficient, conditions that sometimes result in the plants going into physiological check and almost always causing transplants to be slow and difficult to establish.

Failure to give container-grown shrubs sufficient room between plants in the standing-out beds prevents their lateral bud development; it produces poorly shaped plants with bare sides and a tendency to increase in height but not to spread when planted out.

Transplanting during production

Open land to open land

Open-land-to-open-land transplanting accompanied by root pruning and spacing out is a standard beneficial practice in producing both large and small tree transplants. Seedlings destined to be sold as whips have their taproots shortened and laterals induced by either undercutting or lifting and replanting.

Open land to container (containerization)

Moving tree transplants, root-balled conifers and imported evergreens from the open land to containers is commonly practised. Half-hardy species such as olive often have a little foreign land clinging to their roots, as the appearance of alien weeds showed in the planting of the Eden Project! There is nothing wrong with this production method providing that rigorous phytosanitation procedures are followed and the plant's roots have become fully re-established in the container substrate before sale. Most shrubs are now grown in containers throughout their time in the nursery.

Potting on, container to container

Most container-grown stock will have been moved on from a liner to the container in which it is sold; typically for herbaceous types and shrubs this has a 3 or 5 litre capacity. Large shrub specimens may have spent a period in such pots before their final move. Purchasers need be interested in these manoeuvres only if the specimen has failed to root out into the larger pot. This is usually because the plant is offered for sale too soon or was rootbound in the original container.

Formative pruning and training

Much stock, particularly the larger grades, will have been pruned during production. This will have been done to shape the plant by encouraging the production of vigorous shoots known as 'brakes' or to remove unwanted laterals as with standard trees suitable for street-side planting. When transplants of a similar stature but with all their side branches retained are required, the order must explicitly stipulate 'feathered' specimens.

The nursery pruning of shrubs is often influenced by a garden centre customer's concept of quality. Amateur gardeners tend to want transplants that appear as miniature replicas of older specimens. To this end, today's nurseries often shear over shrubs to induce as many breaks (shoots) as possible. In fact many kinds of shrub develop their permanent branchwork from only a limited number of these shoots. In such cases the preferred plant should have a modest number of strong stems.

Nursery storage, packing and dispatch

Batch lifting of field-grown stock under favourable conditions at the start of the dormant season is advantageous to the running of a nursery and should enable well-graded stock in excellent condition to be dispatched at any time throughout the transplanting season. To achieve this clearly requires careful lifting and storage. The publication by the Joint Liaison Committee on Plant Supplies (1981) *Code of Practice for Plant Handling* is a useful guide to what should and should not be done.

Field storage

Nurseries exempt from deep-penetrating frosts may still use headland storage for both bare-root and balled-and-burlapped stock. Set up behind temporary shelter to avoid desiccating winds, this may take the form of either 'heeling-in' or the burial of roots and lower stems beneath straw or woodchips; both approaches are illustrated in Fig. 6.5.

Barn and cold storage

Simple frost- and draught-free barns provide satisfactory short-term storage. The roots of woody plants are protected and kept damp beneath straw. Herbaceous subjects are likely to be boxed up unless already containerized.

Specifically designed stores can be temperature and humidity controlled, enabling dormant plants to remain in good condition for long periods. Some herbaceous subjects are even held frozen in their containers. Keeping cold-stored stock dormant can very successfully extend the field-lifted transplant season by 1–2 months, allowing the work to be carried out under favourable conditions. If the site works programme indicates that planting is going to be late – that is, after the start of the growing season – arrangements should be made with the suppliers for stock to be

(a)

(b)

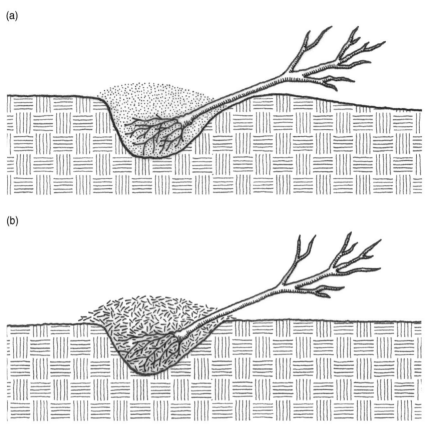

Fig. 6.5. Heeling-in lifted stock. During the dormant season, both bare-root and balled-and-burlapped stock may be held between lifting the plants and planting. Roots are kept moist and frost free by covering with (a) soil or (b) straw, while leaning the plants down reduces exposure to drying wind.

held in suitable stores. Rapid budburst, leafing out and subsequent growth are typical of cold-stored, late-planted stock, making it essential to maintain soil moisture during that critical growth period.

Small container-grown stock is commonly lifted from the standing grounds, checked, graded and packed onto 'Danish trollies'. To protect such stock, many nurseries wrap the trollies in plastic film. This is an excellent practice and protects against mechanical damage, exposure and excessive cold; however, they should receive appropriate attention on arrival on site (see page 109, Arrival on site and temporary storage).

Forms and Types of Transplant Traded

7

Specifications Based on Methods of Production

Stock lifted from open land

Traditionally, trees and shrubs were lifted from the open ground in either autumn/ early winter or early spring and traded as either bare-root or balled-and-burlapped (rootballed) specimens (Fig. 7.1). In both cases the amount of root retained is only a small percentage of the total; plants have, however, what may seem to the novice to be a remarkable ability to regenerate what is in effect a new root system. To some extent this apes nature because woody plants are continually producing and casting off many of their smaller roots. The essential requirement is to prevent the transplant's remaining roots from drying out or suffering physical damage. As noted above, with these conditions fulfilled, stock may be safely lifted and held before dispatch.

Bare-root stock has had the soil removed from around the roots. It is good practice to prune back broken or damaged roots prior to putting them in a moisture-proof bag. Balled-and-burlapped stock has soil retained around the roots. The resulting rootball is wrapped in a loose-weave hessian square, which is tied around the base of the stem. Larger specimens should retain their burlap wrapping when planted; however, it must not be allowed to form a barrier to moisture between the rootball and the surrounding soil. Bare-root, balled-and-burlapped and containerized root systems are shown in Fig. 7.2.

Fig. 7.1. Balled-and-burlapped (rootballed) stock. Stock as delivered from the nursery. Root-balled specimens were planted with the burlap in place. Bare-rooted specimens have their wraps removed. Photo courtesy of Mary Payne.

Stock grown in containers

Most shrubs and many trees are now traded in containers. Their history prior to sale varies; some will have spent their entire production period containerized, others will have been lifted from the open land some time prior to being offered for sale. There are many types of container, ranging from very basic black plastic bags through the standard type of rigid plastic pot to some of more elaborate design. All but the paper tubes noted in the next paragraph must be removed at the time of planting.

Containerized stock is less vulnerable to mistreatment prior to planting, but requires watering throughout this time. Container-grown very small tree transplants are typically 1- or 2-year-old seedlings grown in a range of specifically designed containers. The designs range from biodegradable paper tubes that break down in the soil so need not be removed, to moulded plastic pots. Some, such as those known by the trade name Rootrainers, are ribbed to direct roots downwards and so reduce root spiralling. At the time of planting the containers can be split open to remove the plant.

Stock prepared for sale in large containers

Trees ranging in size from heavy standard to semi-mature are sold in containers. Typically they will have spent the greater part of their life growing in the open

(a)   (b)   (c)

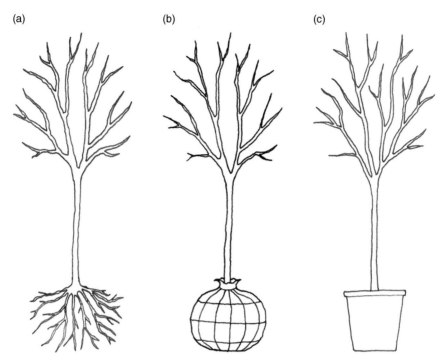

Fig. 7.2. Example nursery root systems. The roots of the vast majority of tree transplants arrive on site in one of three forms: (a) 'bare root'; (b) 'balled and burlapped' also termed rootballed; and (c) containerized. Each gives excellent results when correctly treated.

land before being lifted and containerized. Specimens are ready for transplanting on a landscape site when their root systems are well established in the container and the tree's canopy shows clear signs of healthy growth made after the plant was lifted. Under the right conditions *and with the correct aftercare* such specimens can be planted throughout the year with little if any check to growth. It must be recognized, however, that healthy specimens will have received expert care and frequent irrigation in the nursery. They require a similar level of attention if they are to establish successfully and repay their high cost.

Recent developments include the introduction of new container designs, the use of specific composts and sophisticated irrigation systems that maintain soil moisture at all times. Square containers are considered to help prevent root spiralling but the greatest development has been the 'air pruning pot'. This ingenious design consisting of a wrap of heavy gauge warted plastic is described on page 56 and illustrated in Fig 5.2. Although the design successfully prevents root spirally it should be remembered that the root ball it produces remains as prone to drying out, before and after transplanting, as that produced in a traditional container.

Specifications based on Morphological Forms and their Uses

Trees

When specifying trees both the transplant's morphology and its method of production should be stated. Tree transplants are typically offered in one of the following morphological forms.

Whips

These are young trees or deciduous shrubs, typically produced from seed and ranging in size from as little as 250 mm to the more common 600–1000 mm. Foresters use a single code to indicate age and management. For example a 1+1 transplant is 2 years old and has been lifted and replanted at the end of its first year within the nursery. Whips are typically traded bundled as bare-root transplants. They are most commonly used in the establishment of hedges and massed plantings, as in tree screens and shelter belts.

The establishment and early growth of small transplants is significantly aided by tree shelters, the function of which is explained on page 113).

Clear-stemmed standards

Such trees are nursery-pruned and/or trained to have a clean trunk to a minimum of 2 m. Some specimens are then further pruned to start the development of their main branches (a crown) at that height; others retain their central leader.

Standards are traded in various sizes, calibrated by both height and stem circumference measured in centimetres at 1.2 m above ground level.

The typical transplant for roadsides or urban public open space is the extra-heavy standard of 10–14 cm circumference, nursery-pruned and/or trained with a clean trunk to 3 m. On most urban roadside locations there is a need for significant further height before the development of the full canopy; hence the need for the transplant to have retained its leader. Trees in urban, predominantly hard-landscape sites may have to have a restricted crown spread. In such locations, taxa, species or cultivars with an upright growth-form resulting from narrow branch angles are preferable to those that require shaping by pruning (see below).

Street trees have in the past faced a high degree of stress from several causes, including drought, poor and/or limited volumes of soil, atmospheric pollution, frequent heavy pruning and mechanical damage. Given such hazards it is not surprising that a few stress-tolerant taxa typified by the London plane *Platanus* × *hispanica* and several limes (*Tilia* spp.) dominate European city treescapes. Thanks to a combination of improved city environments and the knowledge and skill of urban arboriculturists, the tree officers, the palette can be greatly expanded – witness the remarkably wide range of species revealed by John Medhurst's work on the trees in London referred to on page 46.

Clear-stemmed standards when used to provide the third dimension in grassed areas, or when emerging from low massed shrubs, may look gaunt and unnatural. As the trees mature, however, their bare trunks become more in balance with their crowns and hence assume a more natural appearance.

Natural feathered standards

These are unpruned specimens with side-shoots retained 'to the ground'. They may or may not have their leader cut back. Typically, feathered standards are used in isolation as standalone or eye-catcher specimens or in conjunction with a low under-planting. This category includes both large-growing 'forest' species and small-growing, so-called decorative trees. The trees in this latter category generally make an additional contribution through one or more seasons of display. Feathered standards are the preferred choice for plantings required to act as a visual screen. When planting feathered specimens it is important to ensure that they receive the best possible post-planting care, as stress typically results in the loss of lower lateral shoots, which in any case is a natural process. The percentage and position of shoots that develop into permanent branches is both site and species dependent.

Multistemmed trees and clumped seedlings

'Multistems' are a variation on the specimen tree. Some species naturally branch close to the ground but most multistemmed specimens result from pruning and training in the nursery; they have several trunks starting from ground level or just above. Large, well-branched specimens of those species that lend themselves to the necessary treatment are available containerized or less commonly ex-open land. Understandably they are expensive.

Clumped seedlings produce a similar appearance; their multiple stems, later to become trunks, result from a bunch of some three to five seedling trees having being planted with their stems touching at ground level. From then on they are treated as one specimen.

As the attraction of both forms of multistemmed tree rests in their branch tracery and bark, these attributes influence which species are grown as eye-catchers or standalone specimens.

Extra-heavy and semi-mature tree transplants

Large trees have been transplanted for more than 200 years. For most of that time they were successfully moved from one location to another on the same estate and hence had their roots out of the ground for only a few hours. The work was supervised and undertaken by the estate's skilled garden staff. We see an echo of such work with the use of the 'tree spade' to relocate specimens within a development site.

The term 'extra-heavy' usually refers to trees with a girth of at least 100 mm. The term 'semi-mature' is reserved for those with a significantly greater girth and age. They are produced in quantity by only a few specialist nurseries. Such specimens are expensive and are usually only considered for very critical locations or as focal points somewhat analogous to living sculptures.

Semi-mature specimens are traded internationally; olives and palms are among the exotic species that are particularly successful. Typically the trees are sourced, prepared, packed and transported by specialist nurseries. Many such companies will, if commissioned, oversee replanting, securing and stabilizing specimens and providing irrigation. They will also give instructions on post-planting care. The majority of such large transplants are delivered to site either containerized or burlapped (Fig. 7.3). If containerized, the work must have been done long enough before for the root system to have re-established within the container. Those with a covered rootball will have had it secured with straps or a wire-netting mesh before wrapping. The mesh should remain in place after planting. The Newman Frame and Newman Tree-Porter were developed for translocation by road. This equipment handles securely bound massive rootballs that can be lifted, transported and craned into pre-prepared sites. An extended period of post-planting care, with an emphasis on the provision of irrigation, is essential and should be committed to by the client during the design process.

Fig. 7.3. Semi-mature tree transplanting. This field-grown root-balled 7 m *Fraxinus ornus* was selected in the nursery by the designer then lifted transported and planted by the supplier using their specialist equipment and knowledge. Photo courtesy of Mary Payne.

Trained and topiary trees

Trees with contrived forms of either their branchwork or their complete canopy have been design features for hundreds of years. In many cases such plantings play a key part in the structure of the design and almost always become strong features in their own right. Trained and topiary forms require many years of skilled training both pre- and post-planting to develop into the mature specimens seen in Europe's cities and great estates. Unless such management is assured, such forms should be avoided.

Pollards, espaliers and pleached forms were traditionally produced *in situ* and it is essential to recognize that as purchased they are works in progress. Although the pruning and training directly concerns the new growth, it determines the form and location of the permanent branchwork.

At the time of planting, weeping and pleached forms need appropriate supports. Shaped specimens are expensive, and designers are strongly advised to inspect and tag any required. Furthermore the quality of the work varies from excellent to those that have been merely pruned and pulled into roughly the right shape shortly before sale.

Topiary forms including the currently popular 'cloud trees' have their overall form determined in the nursery, some are currently produced in Britain, many others imported. Their shapes are often based on wire frameworks but their long-term success still rests on the frequency and skill of their clipping.

Morphological mutations

Many tree species have produced mutant morphological variants such as fastigiate and weeping forms. Fastigiate or narrow-canopied cultivars are often selected to 'fit' into restricted spaces; many, however, lack the grace of a natural tree form. Specifiers should seek out and evaluate mature specimens because many species, although markedly upright in their early years, develop a more spreading form with maturity. Among those frequently planted are *Tilia cordata* 'Greenspire', *Quercus robur* 'Fastigiata Koster' and *Pyrus calleryana* 'Chanticleer'.

Weeping cultivars, that is, those with pendulous branches, are usually grafted onto upright stems of the species to produce 'standards'. Although seen by some as 'unnatural' they are generally admired and hence selected as isolated eye-catchers. Many require training for several years after transplanting to carry a leader to the desired final height; others such as the Weeping beech (*Fagus sylvatica* 'Pendula') have the happy knack of extending their height while having weeping lateral branches.

Tree distribution and spacing within plantings

WHIP-BASED MATRIXES FOR EXTENSIVE AREAS. The spacing of tree seedlings or whips on a large-scale planting should be influenced by its primary function, which on most landscape sites will not be forestry but an amenity in the form of a narrow band of trees for shelter or visual or sound screening. In these cases, the main value will rest in the outer ranks of trees and shrubs that, having sufficient light, retain their lower branches. To have the best effect such 'woodland edges' should be a mix of tree and shrub species, a point stressed by Le Sueur (1951) in *Hedges Shelter-Belts and Screens*, and more recently by Pollock (1984) *Shelter Hedges and Trees*. Shrubs planted within wide belts usually succumb to low light with few surviving more than 10 years.

In all amenity tree plantings there is need to link spacing and long-term management, in particular the thinning programme. Very close planting of 1 or 2 m makes an early showing, but in the absence of skilled and carefully timed thinning soon produces problems. Compare the 1–4 m² allocated to each tree with the typical mature broad-leaf woodland that has trees at some 6 m apart giving each canopy 36 m². No wonder competition for light *within* the planting causes many losses and poor growth. The suppression of side branches produces a planting of beanpoles! Three metres is the minimum distance apart at which unthinned plantings will continue to thrive for around 15 years while retaining some lower branches.

Large-scale shelter belts merge into woodland but by contrast there is often little space for the visual screening commonly required on development sites. To remain functional for many years without pruning, narrow tree screens may be planted as three rows 2.5 m apart, with the transplants at 2–2.5 m between plants. The provision of shelter for hop gardens and orchards shows that trees such as *Alnus rubra*, even when planted as a single row and limited to a spread of 1.5 m, can form a successful screen. Such 'giant hedges' require skilled training and regular pruning, however. Screens up to 4 m high start as a double-row hedge of vigorous shrubs at 1 m spacing

Gapping up (also known as beating up) should not influence spacing if it is under the control of the designer during the defects liability period; thinning after several years is unlikely to be so supervised and it is prudent to specify a wide spacing if its omission seems likely. The practice of including a small percentage of feathered standards in the mix to provide some sense of maturity should not affect the overall spacing.

Tree plantings using more than one taxon generally follow a variation on one of the following approaches:

- Single-species groups of around ten specimens, which after thinning will leave at least three as adjacent neighbours. As noted above the addition of shrub species seldom produces a long-term understory beyond the woodland edge.
- Uniform mixture throughout the planting; this can provide a satisfactory screen, but it is difficult to retain the same mix when thinning.
- One or more permanent tree species dispersed at final spacing between coppiced specimens or short-lived, fast-growing 'fillers' removed when thinning.
- One or more permanent tree species dispersed at the final spacing between shrub species to produce an understory or woodland edge.

Long-term management is simpler if the trees are planted in rows; however, these should avoid being aligned with the most significant viewing point and are far less obvious if set out in very gentle curves as shown in the diagram of Fig. 7.4 and photograph of Fig. 7.5.

Historically trees were commonly planted in a quincux, a layout in which each tree in turn is regarded as the central specimen in a group of five as illustrated in Fig. 7.6.

STREET TREE AND AVENUE SPACING. The spacing of trees to form an avenue is an excellent example of how aesthetics can be both assisted and frustrated by tree biology and arboriculture.

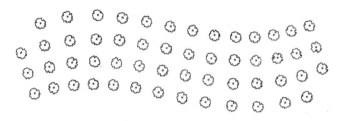

Fig. 7.4. Setting out tree belts. Early year maintenance is easier if trees are planted in rows. To avoid the appearance of a forest plantation the rows should be curved so that from important aspects the planting appears more informal.

Fig. 7.5. Curved alignment of trees. The slight curve on the 3 m wide rows results in an apparently informal layout within this amenity woodland. Photo P. Thoday.

(a)

(b)

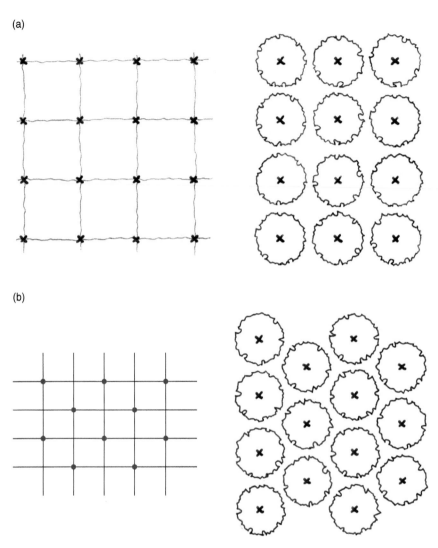

Fig. 7.6. Tree planting grids. Planting on the square is often practiced (a) but the orchard tradition of using the quincunx (b) as the module gives each tree improved all round space for canopy development.

    The designer must decide if the trees are to remain as two ranks of full-canopied 'free-standing' specimens or fuse their canopies. With roadside avenues the road width will determine the distance between the rows of trees but in other locations this may rest with the designer. This offers the opportunity to create the Gothic arch effect seen in the grounds at Versailles where London plane (*Platanus* x *hispanica*) is planted at 9 m between rows. The proximity of one tree to its neighbour produces the acute branch angle. The fifteen year old avenue of American Ash (*Fraxinus americana*) shown on the front cover was planted by the author along the drive to a stud farm in Ontario, Canada. The 7-year-old transplants are spaced at 9 m intervals with 8 m between the two rows.

New tree planting to 'repair' an old avenue is usually a bad decision. Clear-felling and stump removal is a major project but is the only arboriculturally correct solution. Replanting to gap-up losses cannot return the avenue to the essential quality of uniformity. Replacements planted adjacent to the existing trees (or even under their canopy) develop uneven crowns and are unable to compete for light, water and nutrients.

Tree planting along urban streets is often heavily influenced by underground and above-ground services, which usually prevent uniform spacing; such plantings require the agreement of all involved authorities.

Shrubs and climbers

Transplant forms available

NATURAL OR CUT-BACK FORMS. Assessing quality in shrub transplants requires an understanding of the shrubby growth habit. In nature, one shrub growth-form tends to produce vigorous 'water shoots' from old stems near ground level. This canopy regeneration is fundamental to the morphology of such genera as *Spiraea* and *Deutzia* to the extent that the purchased canopy is superseded, its value being in nourishing this shoot production.

In contrast, *Cistus* and *Skimmia* continue to grow from their original branch framework somewhat like miniature trees; hence transplants with this growth form should have sufficient strong breaks to give a vigorous primary branch system (Fig. 7.7).

In the nursery the first shoots from rooted cuttings of shrubs are usually 'pinched back' or 'stopped'. By pruning away the terminal bud lateral buds grow out to produce the first branches. The specimens may then be left to develop naturally or they may receive further 'haircuts' resulting in many more twig-like shoots. Although such 'bushy' specimens may appeal to the amateur gardener, they are not the ideal basis for either of the growth forms noted above. Transplants with a relatively few strong stems are preferable (Fig. 7.8).

BARE-ROOT AND BALLED-AND-BURLAPPED FROM OPEN LAND. Some shrubs, typically those 'tough' kinds that transplant with few losses, are grown on in the open nursery. When large enough they are usually lifted at the start of the dormant season and stored before sale as bare-root transplants. Most are either seedlings or propagated from hardwood cuttings; they are often sold in bundles for use as hedging or planted en masse as woodland edge or understory.

Balled-and-burlapped shrub transplants are less commonly traded than in the past, but evergreens ranging from *Prunus laurocerasus* 'Otto Luyken' to *Rhododendron* cvs are balled and burlapped from the open land as an alternative to container production.

(a)

(b)

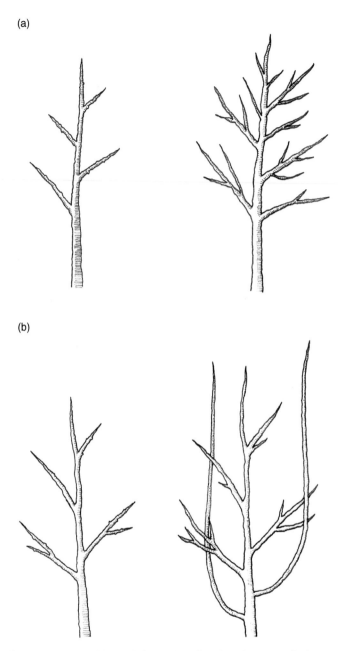

Fig. 7.7. Shrub growth and branch formation. Shrubs show two distinct growth habits. The tree-like form (a) develops by the incremental extension of their twigs. In the other form (b) a succession of vigorous juvenile shoots sprout from near the base or proximal area of the plant.

Fig. 7.8. Shearing young stock. This sequence of three drawings shows overall clipping and the resultant plethora of shoots. Few if any will develop into the shrub's permanent branch structure.

CONTAINER GROWN. The majority of shrubs are sold as container-grown specimens and are offered in a range of pot sizes. The smaller-growing species such as lavender are marketed in 1 litre and 2 litre containers, and it may be possible to buy small transplants known as liners of the larger-growing types in containers of these sizes. The most common containers are of 3 litre and 5 litre capacity, but shrubs may be moved on into 7.5 litre and 10 litre sizes. If available, these larger specimens will be considerably more expensive and, partly depending on species and age, more difficult to establish under landscape site conditions.

Styles and uses of shrub planting

Shrubs are the mainstay of many amenity plantings. Contemporary styles of planting can be summarized in four headings:

- Massed plantings;
- Woody ground cover;
- Emergents from ground cover;
- Individual specimen 'eye-catchers'.

MASSED PLANTINGS. Planting shrubs and sub-shrubs at a density that results in their canopies fusing into a continuous mass within two or three growing seasons has been a popular treatment for more than 50 years. Such plantings may use a single taxon or a patchwork of groups of contrasting form and canopy texture.

All woody plants when planted closer than their mature natural spread will grow to fill the space provided; many – though not all – produce a fused canopy in which individual identity is lost as their stems interlace, ending any further need for weed control or pruning. Interlacing depends on the strong growth of juvenile or vegetative shoots characteristic of young plants. Trials at the University of Bath consistently showed that young vigorous transplants were most successful in producing fused canopies.

The exact spacing depends on growth habit and ultimate size of the selected taxon and ranges from one per square metre for *Prunus laurocerasus* to three for *Symphoricarpos* × *chenaultii* 'Hancock'. The spacing of *Erica* spp. shown in Fig. 7.9 and the ground cover bank of Cotoneaster in Fig. 7.10 were both designed to achieve full cover after two growing seasons.

If planted too close, laterals are supressed and some specimens die while others grow large, producing an irregular effect; with too wide a spacing there can be no fusion. A common error is to start massed shrub plantings too close to a path or bed edge. It is advisable to plant an edge strip of a low ground cover such as *Hedera hibernica* that can be overgrown as the shrub mass expands (Fig. 7.11).

WOODY GROUND COVER. 'Ground cover' is a term used to describe two very different ideas: aesthetic ground cover and biological ground cover. The former describes any low planting that produces the visual effect of the soil being hidden by vegetation. Biological ground cover is where the vegetation and its debris are sufficiently dense all the year round to prevent the growth of other species, in this context weeds. Whereas there are almost limitless options to create aesthetic ground cover, the selection and spacing of the subjects used to produce biological ground cover are critical to its success (Fig. 7.10).

Fig. 7.9. Transplant – optimum spacing. Optimum spacing depends on design intent. For shrub massing and ground cover, full cover after two growing seasons allows the plants time to develop before competing for water, nutrients and light. Photo courtesy of Mary Payne.

Fig. 7.10. Ground cover. A bank of Cotoneaster 'Coral Beauty' three years after planting at 3 per m² from 3 litre containers. Pre-closed-canopy weed control was by woodchip mulch. Photo courtesy of Mary Payne.

Woody plants that produce satisfactory ground cover come in two distinct growth forms; hemispherical as in *Hebe*, and prostrate as in *Juniperus*. The spacing of the hemispherical form requires particular care, too close results in very uneven growth while too wide produces an effect that has been described as 'like a tray of bread rolls'.

Ground cover may be used to provide a short-term solution before being over taken by more permanent subjects. The planting of shrubs to form screens alongside footpaths causes problems if the first row is too near the path edge. As their canopies expand they obstruct the path, necessitating pruning, which creates hard lines that may not be the design intent. As shown in Fig. 7.11, the simple solution is to front the shrub planting with an expendable border of ground cover such as ivy. This provides room for the shrubs to develop thus removing the need for pruning.

EMERGENTS AND 'EYE-CATCHERS'. In Northern Europe historically, shrubby forms were used mainly as 'greenings' – that is, as background or screening vegetation. In the early years of the 20th century, the introduction of shrubs with striking seasonal display heralded the planting of 'specimen' decorative shrubs, a style that is still used in low-maintenance designs.

(a)

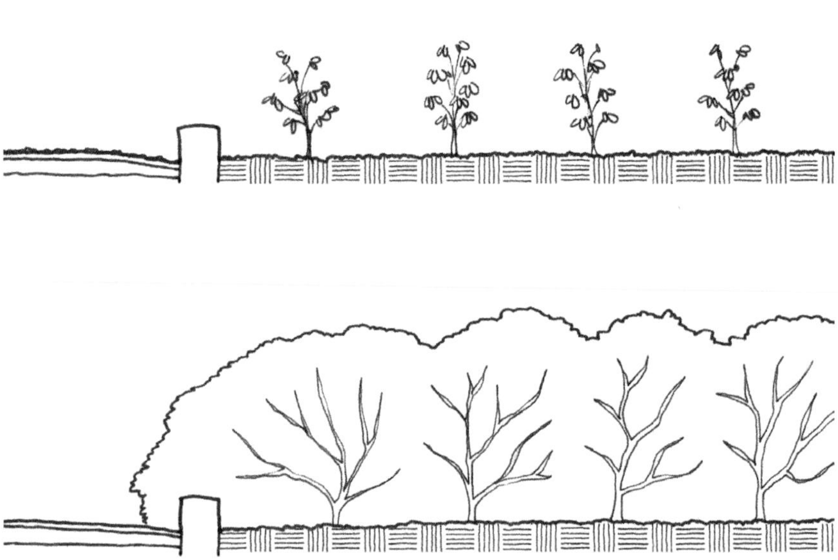

(b)

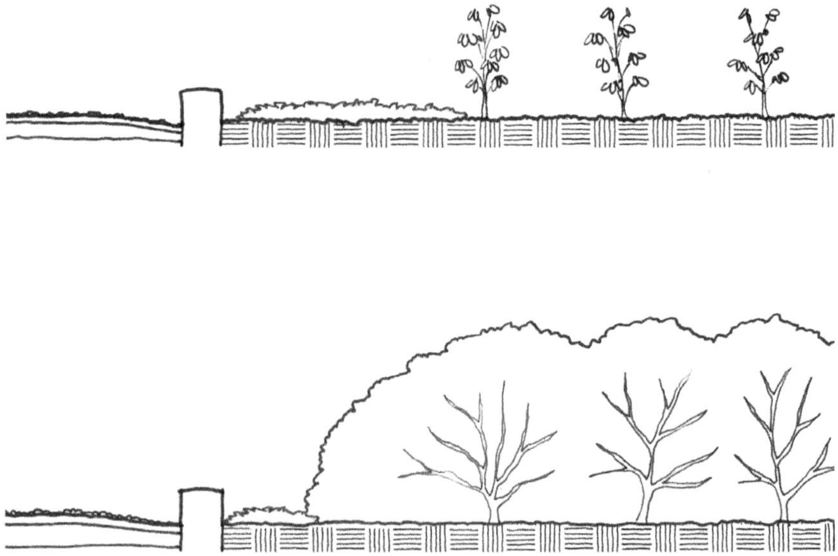

Fig. 7.11. Shrub planting near footpaths. In public landscapes, shrubs overhanging paths are cut back forming a hedge like feature negating the natural form of the design intent (a). A fronting of ground cover provides a short term finish before being overgrown by the shrub canopy (b).

Such outstanding shrubs are often grown as standalone specimens formed from a single plant or a closely spaced group of three. Many are emergents surrounded by low ground cover. Larger transplants are usually specified for such situations.

The usual intention is for eye-catchers and emergents to develop their full potential in size and shape. It is therefore necessary to provide space around each specimen, often by surrounding it with plants that can be removed or overgrown before they cause any biological or visual interference.

Climbers

The climbing growth habit uses several very distinct methods of anchorage, shown in Fig. 7.12. All rely on young growth; hence it is critical that a transplant's new shoots develop *within reach of an appropriate support*. Old stems flop if not supported and large transplants chosen with the intention of making an instant effect often fail to secure the vital first contact.

The so-called self-clinging species such as ivy (*Hedera helix*) and climbing hydrangea (*Hydrangea petiolaris*) produce adventitious roots that grip onto vertical surfaces. Boston ivy (*Parthenocissus tricuspidata*) and some other members of the Vitaceae cling to vertical surfaces by using pads on the ends of their tendrils. Other plants climb by twining tendrils or stems. These require specific supports. Twiners can only use supports of a circumference appropriate to the length and or movement of their climbing organ. Tendril climbers can successfully utilize horizontal supports such as wires; stem-twiners more successfully climb vertical supports.

There is sometimes friction between the planting of climbers and the maintenance of structures. It is pointless to plant perennial climbing plants if they are doomed to be removed during painting and other building upkeep. In such cases it is best to restrict climbers to deliberately planned landscape features such as pergolas and archways. These must be designed with a selected plant in mind to ensure that the location and form of support is appropriate to the plant's mode of climbing.

In addition to method of climbing and growth form, selection must take into account, vigour, ultimate size and weight. These characters vary enormously from the potentially massive *Wisteria* spp. to the wispy growth of large-flowered clematis cultivars.

Many of the plants that clothe the walls of our gardens are, in nature, free-standing shrubs. They do not climb, and have to be trained by tying and pruning if they are to achieve the design intent.

Herbaceous perennials

Transplant forms available and suitability

Traditionally, herbaceous perennial transplants were traded as dormant 'roots' or 'crowns' in the transplanting seasons of autumn–early winter and early spring. The plants were lifted after 1–2 years from open ground with soil around a rootball, which was then wrapped in newspaper before dispatch.

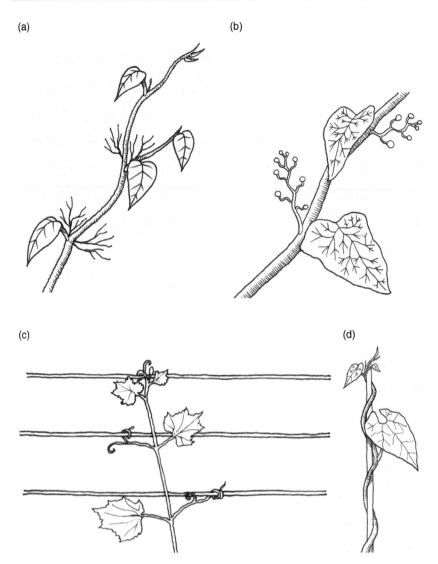

Fig. 7.12. Climbing methods. All climbers need support. Those that produce either adventitious roots (a) or pads (b) can grip on vertical surfaces. Those with tendrils (c) grip onto small diameter objects, whereas those with twining stems (d) spiral round their supports.

Today the majority are grown in 1 litre, 2 litre and 3 litre containers. Transplants from the smaller containers produce only a little flower in the year of planting; however, appropriate types are very successful for mass planting or as ground cover. The largest containers are best left for use at the Chelsea Flower Show! Although container production allows year-round sales, those planted in the summer season when in full growth often make little subsequent growth that year and a rather poor show of flower.

In most out-of-garden schemes there is only limited opportunity to use fully herbaceous perennials because the requirement is usually for the vegetation to present a year-round effect. To achieve all-year-round performance many designers turn to evergreen herbs; the foliage of several such as *Bergenia* cvs and *Helleborus* spp. make successful year-round ground cover.

Appearance before and after flowering is important. Of the many deciduous herbaceous perennials some, such as the day lily, are significantly earlier than others to leaf out so make a contribution to the design in spring. After flowering, the foliage of some such as the garden iris contribute for several months; others 'die ugly' and may be fairly described as a blot on the landscape! As a general rule the later-flowering species such as *Anemone* x *hybrida* avoid this summer downturn.

Some herbaceous perennials can become invasive when planted as part of a mosaic of different taxa. The invaders are usually equipped with ground-level stolons or underground rhizomes. There is nothing inherently wrong with using such plants but they are best planted in isolation. At the other end of the spectrum are those with short lives such as lupins and the high maintenance demanders like the delphinium cultivars; neither group is suitable for landscape planting.

Use

The classic herbaceous border of the early years of the 20th century is very rarely appropriate for the kind of landscapes considered in this book. Their successful use in private gardens depends on their positioning, careful management and demanding seasonal maintenance. They also require a programme of time-consuming refurbishment every few years.

There is, however, a renewed interest in planting herbaceous perennials in public landscapes as is clear from the work of such designers as Mary Payne, Pete Oudolf and James Hitchmough. These exponents are plants-people with an outstanding knowledge of plants, their ecology and their cultivation. Their styles, although clearly different from one another, have several points in common.

All use taxa that are robust and durable, and when combined provide a long season of interest. All have far simpler maintenance schedules and workloads than the classic herbaceous border and none of the plants need staking. Most are established from small transplants arranged in groups of either a single taxon or a mix of several. All the designs require meticulous attention to the elimination of perennial weeds either before or during land preparation. Post-planting, many of the designs benefit from mulch, irrigation and nutrition during their establishment, but the long-term nutrient level of the soil tends to be lower than that for traditional herbaceous borders. Many of the designs can be cut to the ground once per year either in the fall or early spring if dead stems, leaves and seed heads are a seasonal feature. A photograph of Mary Payne's low maintenance year round interest design is shown in Fig. 7.13. One of James Hitchmough's unique designs for the Olympic Park in London is shown in Fig. 4.1.

Fig. 7.13. Low-maintenance planting. Herbaceous perennials including grasses are inter-planted with coppiced willows and Cornus to produce seasonal change and interest. The design is cut over once in late winter. Photo courtesy of Mary Payne.

The creation of species-rich meadows is addressed on page 103, in the section on Establishing native perennial herbs.

Bedding plants

For some 150 years, flower bed and container seasonal colour has been based on 'bedding plants' – that is, plants treated as single-season plantings raised in nurseries and planted out prior to flowering. Some are true annuals, others either biennials or perennials treated as annuals. There are literally thousands of cultivars of dozens of species offered for use in private gardens but only a small number are satisfactory for public landscapes. There is little recent literature on this subject as practiced professionally; however, Raymond Evison's (1958) *Gardening for Display* remains a valuable guide for those concerned with public open space.

In spite of changes in landscape and garden fashion, bedding plants remain the main source of floral colour within urban areas (Figs 7.14 and 7.15). The landscape designer's primary concern is likely to be the installation of appropriate beds and raised planters; however, their first planting may fall within the scope of the capital works.

Fig. 7.14. Seasonal urban green-space planting. A simple bed shape simplifies maintenance while providing colour in the city centre of San Sebastian. Photo courtesy of Mikel Pagola.

With the exception of wallflowers lifted bare-rooted from the open land, bedding plant transplants are glasshouse-raised from either seed or cuttings and sold in containers; the cost per plant is influenced both by the kind of plant and by the type of container used. Container design determines the glasshouse space and the volume of compost each plant has. 'Strips' are plastic troughs holding a single row of typically five transplants. As they share the same root-run they have to be pulled apart before planting. More costly, but higher-quality transplants are sold individually or in isolated cells within multipacks each of at least 350 cc capacity (7 cm or 9 cm diameter containers). The largest transplants, such as zonal pelargoniums, begonias and fuchsias, may be available in 11 cm or 1 litre pots. Pale foliage and etiolated stems are generally indications of poor quality, suggesting that plants have been grown in low light or at too high a temperature, or in nutrient-deficient compost or a combination of such conditions.

Use

Bedding plants are sold during the transplant seasons March–June for summer flowering taxa and October–November and again in March for spring-flowering types. For summer flowering it is important to select those with a long flowering season in order to carry the display through well into October. In view of the

Fig. 7.15. Inner-city floral display. The provision of appropriate street furniture allows for both supported and free-standing seasonal floral colour. Photo P. Thoday.

range of climatic conditions experienced in many summers, both the plants and their flowers must be able to withstand seasons wet or dry and hot or cold, although each kind will have its optimal set of conditions. Spring-flowering bedding cultivars may give a long season of display if the selection is made from among those that start flowering in late winter.

Bulbs and corms

Dry

Bulbs are composed of fleshy leaf bases, corms are truncated stems; both are perennating organs carrying the plant through its dormant season. Spring-flowering bulbs are available dry from specialist wholesale traders in Britain only between late August and November. Bulbs are graded by size or in the case of *Narcissus* by either size or the number of independent bulbs or 'noses' held together by an outer layer of dried scales.

Traders produce descriptive illustrated catalogues that also define any trade terms used. The majority of the *Narcissus* are British grown; most of the other types are imported.

Of the many genera listed in bulb catalogues only a few are useful in landscapes outside the garden. Their selection is linked to their role in the design. The usual planting style associated with new developments creates large drifts of the same taxon as a permanent spring feature within a grass sward. The degree to which this may be described as 'naturalistic' depends on both the taxa planted and the pattern and spacing of the planting (Figs 7.16 and 7.17).

In-growth (green)

Although snowdrops (*Galanthus* spp.) are a common 'wild' plant in many parts of Britain, none is now regarded as truly native. Snowdrops as a group are not easy to use in new designs. Huge numbers are needed to make any impact. In addition there is the difficulty of obtaining bulbs in good condition; once out of the soil they dry out very quickly and should be planted within a few weeks of lifting. A traditional way of overcoming this is to plant 'in the green', i.e. between flowering and dying down. This is expensive and limited to a brief period each year; it is unnecessary if freshly lifted bulbs can be obtained.

Uses

BULBS IN TURF. Some would say that there are already quite enough roadside bulb plantings based on crowded ranks of a handful of low-cost narcissus cultivars originally mass-produced for cut-flower production. Regardless of this

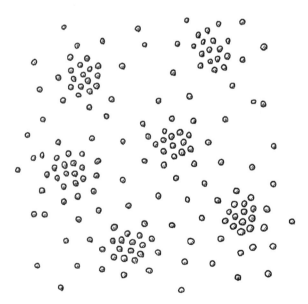

Fig. 7.16. Bulb distribution. Planting bulbs in groups produces a natural effect. In the wild many bulb forming species occur as scattered clumps each originating from an individual bulb. Seedlings play an important part in the formation of continuous sheets such as are found in bluebell woods.

Fig. 7.17. Naturalistic bulb planting. Recently planted crocus distributed to give the appearance of a naturally occurring population. Photo courtesy of Mary Payne.

possible overuse, when skilfully selected and arranged bulbs remain irreplaceable harbingers of spring. Bulbs for grass areas must be selected for both their appearance when in flower and their persistence. Cultivars of *Narcissus* are by far the most successful in either uncut or rough-cut grass providing the latter sites are not mown for 6 weeks after the flowers fade. Crocus establish well in more frequently cut areas if the 6-week rule is observed.

BULBS IN SEASONAL BEDDING. Bulbs used in such a way are lifted and discarded after flowering, so their role is similar to other bedding plants. If horticultural site works and the proposed handover date coincide with bulb planting and flowering, bulbs will provide a welcome early display of colour in an otherwise immature design. Tulips are generally the most successful of the bulbs, with early- and late-flowering cultivars available among the many offered by specialist bulb wholesalers.

BULBS AND CORMS IN WOODLAND AND WOODLAND EDGE. Four British native/naturalized bulbs and one tuber – namely bluebell (*Hyacinthoides non-scripta*), wild daffodil (*Narcissus pseudonarcissus*), wild garlic (*Allium ursinum*), common snowdrop (*Galanthus nivalis*) and wood anemone (*Anemone nemorosa*) – produce memorable sights beneath deciduous woodland when found in sufficient numbers. While it is all but impossible to plant enough to give such an effect in the short term, the nucleus of a population can still be established by planting one or two clumps of bulbs per square metre. For bluebells, seeding is a perfectly effective technique but requires specialist site-specific advice, for example the National Wildflower Centre at: www.landlife.org.uk.

Turf

A wide range of turf is traded; it ranges from that cut from meadowland to that raised in turf nurseries using selected mixes of named grass species or their cultivars. Turf suitable for general-amenity areas is field grown, while much of the highest 'lawn' grade is produced on specific substrates spread on plastic sheeting. For prestigious 'lawn' areas the species/cultivar composition of the sward should be specified.

Seeds and Direct Seeding

8

Landscape designers are rarely directly involved in the purchase of seeds; however, seeding on site is used to establish areas of turf of various grades of fineness, and for uses ranging from 'look-upon' to 'walk-upon' to 'play-upon' to species-rich grassland. Direct seeding is also used to produce displays of native and exotic annuals, whereas the on-site seeding of trees and shrubs can be used to establish woodland.

Some understanding of the seed biology of the groups of plants specified in direct seeding will help in the appreciation of some of the issues inherent in their use.

The purchaser of crop seeds is safeguarded by the Seed Acts, which are designed to ensure the purity and vitality of seeds offered for sale. Unfortunately they do not apply to decorative species; however, reputable companies issue the date of harvest, the percentage germination and where appropriate, as when used on conservation sites, the seed origin. The quality of the seed of forest tree species is safeguarded by the regulations on forest reproductive material (which also includes cones, cuttings and transplants). For more information, see Forestry Commission (2002) *Forest Reproductive Material (Great Britain Regulations)*.

Wild species of both herbaceous and woody plants often have one of several complex germination requirements which their cultivated forms have had bred out of them. Seeds having these survival strategies may require some environmental stimulus before they can germinate; this can result in delayed or staggered germination and in some cases seeds remaining dormant in the soil for a year.

By contrast, seeds of 'recalcitrant' tree species have a very short life in storage and may need to be sown within weeks of harvest. This obviously narrows the window of opportunity in any one year.

Turf Grass from Seed

The amenity grass sward

The amenity grass sward has a long and interesting history in which its composition, management and use have always interacted.

Within that history are to be found medieval flowery meads, the scythed quads of Oxford and Cambridge colleges, the playing fields of Eton, the keep-off-the-grass pony-power-mown lawns of Victorian parks, Edwardian bowling greens and today's return to an appreciation of species-rich grassland.

The successful establishment of many of today's grassed features requires specialist advice and specialist preparation, establishment and upkeep; however, the basic amenity turf that sits at the heart of countless areas of public and institutional greenspace is rightly regarded as part of the general undertaking in the planting of such sites.

The British climate favours such basic green swards; grass grows for nine or more months of the year and neither our winter chill nor a summer drought is severe enough to prevent it from remaining green throughout most years.

It is the very ease with which grasses grow that allows a cavalier approach to the establishment of turfed areas. When pushed to extremes, however, the 'grass grows whatever' approach results in a scruffy appearance that both destroys the image of a uniform sward intended to frame the plantings and fails to withstand the wear of feet and informal play. Preparing the site, selecting the most suitable grasses and establishing the sward are all of equal importance (see page 124, on Turf laying, and Direct seeding of amenity turf, page 125). The development of larger, more robust mowing machines has made the frequent mowing of large areas of general-amenity turf economically feasible, resulting in a denser, more uniform sward.

Turf grasses

The history of the selection of grasses for amenity sites is one of increasing sophistication. It started in the 18th century with the use of 'hay seed' swept from barns that had previously stored hay cut from long-established meadows. The early deliberate seed production of selected species led to the era of the *Agrostis/Festuca* mixtures. Over the past 50 years, plant-breeders have added amenity grass cultivars to those bred for agriculture. Although the focus of attention has been on cultivars for sports surfaces, very helpful recommendations are now available from the specialist seed houses for general-amenity areas tailored to suit both site and use.

The standard practice is to sow a mix of cultivars derived from several species. This approach combines such qualities as winter green, denseness of sward, hard wear and short stature. The most spectacular and useful development has been in the breeding of dwarf but robust cultivars of perennial rye grass (*Lolium perenne*) to produce excellent amenity swards.

On areas, particularly brownfield sites, identified as general public open space, seed mixes should include a nitrogen-fixing legume such as wild white clover (*Trifolium repens*).

Seed or turf?

The comparative advantages and disadvantages of seed and turf may be summarized under two headings: short-term effect and cost. Turf skilfully laid can produce an instant effect at almost any time of the year; it can be walked on without damage after 4 weeks. Seed is season dependent; under good growing conditions it produces a green haze after 4 weeks, a complete cover after 8 weeks and can be walked on after 10 weeks.

Both turf and seed require keeping moist during their establishment period – seed to germinate uniformly, and turf to root into the substrate and not shrink at the edges. A standard mowing regime, if necessary and appropriate to the season, may be started on turf after 3 weeks. Cut seedling grass when 10–15cms high, depending on season this will also control annual weed seedlings. Typically routine mowing may start two weeks later.

Taking into account site preparation, good-quality materials and all labour, the cost of a sward from seed is typically around 10% of one from turf. In addition to cost and speed, matching grass species/cultivars to site conditions and future use is easier with seed; it is not uncommon for inappropriate grasses in otherwise high-quality expensive turf to die out after 2 or 3 years.

Sports surfaces are not considered in this book; see Christians, N.E. (2011) *Fundamentals of Turf Grass Management* and Emmons, R.D. and Rossi, F. (2015) *Turfgrass Science and Management*.

Direct seeding of woody plants

There is long history of establishing stands of trees by 'natural regeneration' – that is, by allowing seed from nearby trees to colonize sites. *In situ* or direct seeding of areas received a boost of interest in the 1970s. Forty years later, sites tilled and sown with the seed of selected trees and shrubs are supporting fine belts of low-cost young trees. Such direct seeding fits well with both the use of local genetic material and the concept of 'advanced planting' (see page 37).

That direct seeding remains fairly uncommon appears to be due to a combination of factors. At the top of the list of perceived disadvantages is the desire to rapidly achieve a visual 'statement', if not a sense of maturity. Depending on the species, the following gives some indication of the time needed to achieve an impact: seedlings reach between 50 and 150 mm in their first year; this increases to between 100 and 400 mm by the end of the second year; and from then on growth rates should be similar to whips of the same species.

Slow development can be mitigated by interplanting a small number of feathered standards, say 25 per hectare. Other potential problems include the occasional shortage of good-quality seed of the chosen species and the risk on some sites of seed and seedling predation. While Forestry Commission reports give only guarded approval, we might contrast this with the high performance achieved in tree nurseries. With all parties fully committed and recognizing the timescale and the care involved, direct seeding remains a valid option for appropriate amenity landscape sites.

To a considerable extent, tree seed biology influences both species selection and sowing times. The seed of some tree species is produced irregularly, in only what are known as 'mast' years, making its availability uncertain or dependant on carefully controlled storage.

The second biological intervention is linked to the seed's shelf life. Recalcitrant seeds cannot be either dry- or cold-stored and must be sown in the autumn they are shed. Storable species can be spring sown.

Direct tree-seeding requires forward planning. The undertaking involves land preparation and seed availability, its storage (or non-storage), sowing rates, seed dressings and sowing techniques. Advice on all these matters is available from specialist suppliers and/or contractors.

Weed cover can be very detrimental; hence control is important, the best method being linked to the dominant species; in many cases the control programme starts with a pre-emergent herbicide application. Very successful stands will need thinning in their second year.

Direct Seeding of Native and Exotic Annuals

Very successful low-cost seasonal displays can be achieved using directly sown native and exotic subjects. The 'annual border' is a rarely seen old style of floral display produced by the direct sowing of hardy annuals, resulting in an effect rather in the style of a Gertrude Jekyll herbaceous border. Recently two approaches to the direct seeding of annuals have given very effective displays. In one case using native wildflowers and in the other exotic species and their cultivars, both have proved to be suitable in some public landscape settings. Depending on the mixture, annuals come into flower by midsummer for a period of some 3 months. Some remarkable and very colourful results have been achieved on apparently unpromising sites (Figs 8.1 and 8.2).

Native arable land weeds

Arable-land weeds once common in our cornfields have long been regarded as among the showiest of our 'wild flowers'. Corn poppy (*Papaver rhoeas*), cornflower (*Centaurea cyanus*) and corn marigold (*Chrysanthemum segetum*) are in fact early, uninvited migrants imported with crop seeds.

Selected mixes are sown either on a prepared clean seedbed alone or mixed with a short-strawed wheat to produce an 'old-fashioned cornfield' effect. Many are winter annuals: they germinate in the autumn, over-winter as seedlings and

Fig. 8.1. Direct seeded wildflowers. One of many successful brown-field urban sites carried out by Landlife using native species. Photo courtesy of Richard Scott.

flower next summer; others germinate in the spring and flower 3 months or so later. If, after flowering, areas are left the mature seed may fall to form the beginnings of a seed bank, but it is very unlikely that without some surface cultivation a permanent, year-after-year display can be established this way.

Seed of appropriate species is available in sufficient quantity to sow large areas. The National Wildlife Centre notes that such sowings may be done in either autumn or spring following a surface cultivation to establish a seedbed. Reputable suppliers provide details of seed origin, advice on seed mixes, sowing rates and cultural details.

Exotic annual species and their cultivars

These when sown in late spring produce outstanding displays, flowering from June to September. Displays of exotic annuals use a skilfully blended mix of species and cultivars from around the world. Unlike with most planting concepts, a great deal of research, such as that done by Professor Nigel Dunnett of Sheffield University, has gone into refining this topic and it is very advisable to liaise with specialists for husbandry advice and to buy already mixed blends to suit the location and deliver the design intent.

Fig. 8.2. Direct seeded exotic annuals. Selected mixes of annuals provide a succession of flowers for several summer months. Photo courtesy of Mary Payne.

Establishing Native Perennial Herbs

Today's interest in establishing native perennials is most often as the essential component of what have become known as species-rich meadows, although the term covers habitats ranging from grasslands to heath and woodland. Although such diverse habitats require site-specific selection of suitable species, low-fertility soils have proved to be the common feature of many long-term successes.

The subject has received a great deal of coverage in the media but the many unsuccessful attempts to establish such habitats makes it clear that a cavalier approach gives poor results. Considerable research has now provided us with a good understanding of appropriate species, their seed biology and the husbandry required to establish them in locations ranging from urban brownfield 'waste' to large nature reserves in the countryside. Very experienced and soundly based information is obtainable from Landlife at the National Wildflower Centre. The Centre has also produced some excellent guidelines, such as Grant Luscombe and Richard Scott (2004) *Wildflowers Work*.

Such species-rich features may be established by transplanting small seedlings, known as plugs, by direct seeding or by a combination of both methods. Seedling size when planted out is based on the container and varies from 1 cm cubes of compost known as plugs to the 9 cm-diameter multipacks used for bedding plant transplants. Such transplants may be used to establish a mixed

herbage on bare soil or planted into an existing sward. In the latter case, establishment is often difficult and should be done in early spring before the growth
of both the grass and the transplants has started, thereby reducing the risk of the
transplants drying out.

Direct seeding

The establishment of species-rich ground floras by direct seeding has been exten-
sively studied; the effect intended ranges from heather moorland to a woodland
ground layer. The most common objective is a durable, species-rich sward in which
the colourful flowering species, the forbs, become a permanent and significant
component. Seed of some 50 species is available. As most herbaceous perennials
do not flower until their second year, grassland mixes usually contain some annuals
to provide colour in the first year. Contrary to what one might think, experienced
practitioners, including Landlife, recommend sowing a 'forbs-only' mix and allowing the grasses to come in naturally; however, there are 20% forbs to 80% grass
mixes available but they should only be used where there is a high risk of erosion
or very heavy early wear. Over-seeding usually by slit seeding following very close
mowing can produce satisfactory results with vigorous species that can compete
with the grass. Small seeded and less vigorous species are unlikely to establish.

Schemes based on establishing a community of only native species will require seed from stock of known origin. Such communities can be sustained by
natural regeneration; to achieve this, areas must be left uncut for a period after
flowering so that mature seed may fall, to form the beginnings of a seed bank.

This is a challenging undertaking, with many examples of failure due to lack
of commitment; specialist advice is strongly recommended because the best approach varies from site to site.

Site Work Before Planting

9

Preceding sections have considered the nature, selection and use of the various categories of plants commonly used in landscape schemes. This section is concerned with the horticultural issues and tasks resulting from their use. The practices discussed have a direct impact on their establishment and their ability to achieve their full potential in delivering the design intent.

On-site horticultural work can gain much from the approach taken by foresters, commercial nurseries and crop-producers. None of these groups spend more effort and time than they need at any stage in the husbandry of their 'crops' but, conversely, *they do not omit to carry out the critical minimum inputs to the required standards at the correct time and under appropriate weather conditions.* Meeting these objectives and working within their constraints is not easy on a landscape project, but pretending that they can be ignored generally leads to failure.

Timing Operations

Horticultural contractors and supervising officers may wish that soft-landscape works could always take place in the most favourable season and after the external building activities have ended but this is not always possible within a complex construction programme. Concurrent building and soft-landscape works can lead to problems of access and damage to both soiled and planted areas. In practice no matter how skilled the compiler, timetables can be frustrated. Ideally the work programme, including obtaining stock, site preparation and horticultural works, should ensure that tasks are undertaken in the right sequence and in the optimum season; but achieving this while harmonizing with building programmes and construction contracts is neither easy nor certain to succeed.

Programme planning and seasonal operations

A well-planned agronomic work programme identifies the labour, equipment, materials and plants needed, and estimates the time required to carry out the work. Ideally, the placing of this programme within the calendar should be based on the influence the seasons have on biological functions such as root growth and bud-burst – as land-work has always been. In this ideal world, even the calendar gives way day on day to the prevailing weather conditions.

The traditional transplant season across the temperate world was split by the winter into two periods, both falling within most plants' dormant period. With a slight adjustment for local conditions, the first period, autumn planting, ran from late October to mid-December, and spring planting from mid-February to the end of March. For most plants originating from a cool temperate climate, the autumn option has the advantage of some immediate root activity that will start the establishment process. If, however, the subjects come from warmer climes it is better to delay planting until the spring, thereby allowing the plants to over-winter in the care of the nursery. This long-established timetable still has relevance; in spite of all the advances in the science and practice of agronomy, it still holds good and is not preferred out of sentiment but because it gives the best results.

Life on a building site is driven by other considerations and the horticultural programme, usually referred to as 'the soft landscape works', must fit within the master timetable. It must start and be completed as scheduled, although rigid adherence to the timetable regardless of weather and site conditions may endanger the outcome. The work may carry on regardless, but neither the soil nor the plants are unaffected when conditions are dismissed as being of little consequence. In such a bleak scenario soil handling and tillage are the most common cause of long-term problems (see the section on Soil spreading and tillage, page 108).

In some projects, programme slippage may place part of the work 'out of season'. In other projects, the demands of timetables or the impatience of owners result in the whole of the soft-landscape contract taking place outside the traditional transplant period. Transplanting can be successfully carried out throughout the year, although 'throughout' should not be taken to mean 'on any day'. Weather conditions such as frost and torrential rain should still have a veto on the day-to-day timetable.

The use of container-grown stock, careful stock storage, irrigation, mulching and covering topsoil stores all help to counter adverse seasonal weather conditions. Planting in the growing season can give excellent results and it is then that soil conditions on development sites are often most amenable. Irrigation must be available, particularly for use during the spring flush of soft growth when both herbaceous and woody plants are most sensitive to drying out. Table 9.1 summarizes the concerns typically encountered in each of the five periods of the 'transplant year'.

Table 9.1. Planting seasons.

Season	Advantages	Disadvantages
Autumn/early winter	Active roots. Good soil temperature.	Risk of waterlogged soil.
Winter	Soil moist.	Risk of frost damage. Low soil temperatures. Inactive roots. Desiccation of evergreens.
Early spring	Root activity before shoot growth. Good soil moisture.	Risk of prolonged winter conditions.
Late spring	Root activity. Satisfactory soil temperature.	Dry soil. Risk of desiccation of new growth.
Summer	None	Dry soil. Excessive air temperature.

Pre-cultivation Tasks

Soil survey

An assessment of the soil to determine texture, structure and the depth of topsoil and subsoil (discussed in Chapter 2) is vital and should be done before the site is disturbed. Such information will indicate how the soil will respond to stripping, storage and subsequent spreading, and determine any need for additives.

Pre-tillage weed control

As has been emphasized throughout this book, the most cost-effective and successful time to control weeds is at a project's inception. Weeds growing on future planting sites should be killed by an application of a translocated herbicide before any soil disturbance covers and shields them. It is important to follow the maker's instructions to ensure sufficient time is left to allow the chemical to break down before any soil tillage or planting takes place. Long-running projects may have topsoil storage heaps; these should be kept weed-free to prevent the build-up of weed seeds in the soil.

Resident pests and diseases

Sites previously cultivated or covered with vegetation may carry a population of phytophagous species such as slugs and vine weevil. As suggested on page 15, the

Pests and diseases section, it is usually easier to control such populations before building works start. (Old tree-stumps may carry several fungal pathogens – hence they should be excavated and removed before cultivation risks spreading infection.)

Cultivations Before and at the Time of Planting

Soil spreading and tillage

In many cases planting areas have to be resoiled to bring them to the required depths or levels. Two factors require particular consideration: the condition of the material being covered and the nature and handling of the transported material on page 27, the Treatment of damaged soils section. The temptation to cover (blind over) denatured or compacted areas must be resisted. Buried material can only become a subsoil that can be utilized by roots if loosened prior to soil spreading.

It is common practice on large sites to create contoured mounds both for their visual value and to save the cost of removing material from site. The resultant mounds are usually built, and consolidated against slippage, by heavy machinery. Determining bulk density and hence the degree of consolidation required rests with the structural engineer and depends on the function and angle of the slope. Given a topsoil depth of 600 mm, neither the nature nor the density of the core material is of great horticultural consequence. In some cases the topsoil is spread after consolidation, in others both the core material and the topsoil are consolidated and then the topsoil is eased to the depth required for planting. The latter approach is considered to achieve a better bonding of core and soil on slopes but risks severely denaturing the topsoil.

Soil inversion must be the most unusual form of tillage used in landscape work. It uses heavy equipment to very deep-plough, and so bury, the (over) fertile and weed-seed-rich topsoil. This brings subsoil to the surface to form a suitable seedbed for native herbaceous species. This technique is fully explained by Luscombe, Scott and Young (2008) in the Landlife article on *Soil Inversion Works*.

Soil levels and edges to hard landscape

The edges of hard-landscape features should determine soil levels. Curb haunchings should fall away at a steep angle to provide space for a sufficient depth of soil to prevent turf drying out. Exposed haunchings are characteristic of substandard unsupervised work. If there is insufficient topsoil available on site to provide the required depths or levels, more must be brought in. Allowance should be made on both turf and cultivated areas for settlement. Lumpy newly spread soil settles as it breaks down into a finer tilth.

Turfed areas are particularly sensitive to edge-level finishes. Traditionally, good garden practice aimed to finish the turf approximately 4 cm above any adjacent hard surface. Elegant as this is, it is impractical for maintaining turf edges unless supported by steel strip. On most amenity sites the grass should finish flush with the hard surface or curb top. A shortage of soil risks mowers being damaged on hard edges. Flowerbeds were traditionally mounded above their surrounds; today, however, bed profiles usually run to the falls and levels of the surrounding area.

Form and depth of tillage

Easing compaction and reinstating drainage fissures within the subsoil is addressed on page 23, Easing compaction section. On most development sites the larger areas of soft landscape can be fenced off at the outset of site works but those that have not been so protected may need ripping.

Beds and borders are most difficult to cultivate mechanically and are particularly challenging when they contain service runs. Rotary cultivation risks being too shallow and can easily destroy any crumb structure the soil may have retained. These patches of soil are designated to receive the scheme's woody and herbaceous plantings upon which the success of the design rests, yet their cultivation is often both minimal and slipshod. Pre-planting cultivation must break up any hardpan to ensure water, air and root access to the subsoil; it must achieve the incorporation of any additives, as noted on page 28 in the Soil additives section, to produce a friable planting site.

The traditional and still excellent form of cultivation is to hand-dig such areas to a depth of 250 mm with a fork, thereby breaking the boundary between the soil layers and producing whatever degree of tilth is possible. If this method is selected it must be stated in the contract so that it can be fairly costed in the quote. Service runs permitting, skilled operators can 'dig' successfully with a mini-digger or backhoe fitted with a narrow bucket, although some tidying up to reinstate running levels and surfaces will be required.

Pre-planting Plant Care

Arrival on site and temporary storage

Many good-quality plants leave the nursery only to deteriorate between arriving on site and being planted. If contractors are employed then unplanted stock remains their responsibility. To this end, they may require space to construct a secure, sheltered holding compound. It is in everyone's interest that the stock remains in good condition before as well as after planting.

The use of Danish trollies to transport container-grown stock, noted page 71, Barn and cold storage section, is an excellent practice, but if wrapped trollies are left in the sun they can reach damaging temperatures. Closely packed consignments risk becoming enveloped in a saturated hot atmosphere that can trigger outbreaks of leaf pathogens and leaf drop. If a consignment has overheated it should be stood in the shade and sprayed over with cold water. Conversely, in frosty weather the plastic wrap should be retained. The roots of many species of containerized stock are more prone to frost damage than are their aerial parts. For short periods of severe weather, stock can be lightly covered with straw. Long-term winter storage on site requires the management outlined on page 71, Field storage section. Transport and packing may result in plants being in darkness; hence the need to unpack in-leaf specimens as soon as possible after arrival on site.

On arrival, a transplant is living on the water it brings with it. With containerized or balled-and-burlapped stock this moisture is held within the substrate. Typically, containerized stock is grown at around field capacity but such plants are often shipped barely moist so as to reduce weight and avoid water seepage. On arrival such stock should be stood upright with sufficient space around it for air to circulate and then thoroughly watered. In summer, vigorously growing containerized specimens will require daily irrigation. Much container-grown stock is damaged or killed by staff members who, unused to pot watering, merely dampen the soil surface by overhead spraying. To wet the rootball, containers must be filled to the brim.

Balled-and-burlapped stock should be held in a sheltered place, either heeled in or with their roots well buried under straw or woodchips or within a soil heap. This is particularly important with evergreens because they desiccate through foliar transpiration. Bare-root stock has only the moisture within its remaining roots. The immediate concern must be to keep exposed roots moist. Rehydration by plunging into water for a few hours is valuable, providing the roots are still alive; getting dead wood to swell up is of no use, as Dr Richmond Dutton has remarked to me.

Plants that have been grown or over-wintered in polytunnels may have started into early spring growth that will be 'soft' and vulnerable to frost and wind damage. As recommended on page 62, Soft growth section, such plants should have been acclimatized, acclimated or, in traditional gardening parlance, hardened off before leaving the nursery, but if delivered straight from a polytunnel it is good practice to stand them in the storage compound for 2 weeks before planting; there they can be covered with fleece at night if a frost is forecast. As already noted, Joint Liaison Committee on Plant Supplies (1981) *Code of Practice for Plant Handling* is a helpful publication covering many aspects of the topic.

Transplanting and Seeding

10

General Principles Applicable to all Locations and Transplants

It might be hoped that there was no need for any comments on this basic task carried out on almost every landscape site, but there are many failures through poor planting. Two common circumstances determine the approach to transplanting on amenity sites. These are planting into beds that have been cultivated overall and pit planting into otherwise undisturbed sites.

Bed cultivation in preparation for planting should ensure that:

- There are no subterranean hardpans or compacted areas that might impede either drainage or root spread.
- The tilth must be fine and deep enough to allow the planting of stock from containers of up to 5 litre capacity without further tillage.
- All additions, such as soil ameliorants or organic matter if needed, should have been incorporated during bed tillage.

The term 'pit planting' is usually reserved for planting within hard landscape or undisturbed turf/grassland and uncultivated areas. It is also used to indicate that extra excavation is required in cultivated beds if the specimen's rootball is greater than the depth of previous tillage.

The planting hole or pit must be at least 300 mm wider than the rootball and deep enough to ensure that the soil level around the specimen is the same as the one that the plants had previously experienced. The excavation of the pit provides the opportunity to ease compaction, ensure satisfactory drainage and, where appropriate, install irrigation.

The spacing and density of the plants should be indicated by the designer because it will be governed by the design intent. The precise location of any specimens intended to grow to a significant height is a critical part of the design; hence their exact location should have been indicated on the planting plan.

Pre-planting checks

Pre-planting checks should ensure that:

- All containerized stock has been well watered before planting. Sufficient time must be allowed for the water to soak the rootball and for the excess water to drain away.
- Bare-root stock has had broken or badly damaged roots and shoots cut back.
- The soil used as backfill is friable.
- Planting holes are larger than the spread of roots/rootball to allow soil to be placed and firmed around the specimens. The space between the roots and the side of the pit will be governed to some extent by conditions on site and the size of the transplant but should be never less than 150 mm.

Planting checks

Planting checks should ensure that:

- Containers and plastic or burlap wrappings on bare-root specimens are removed. Burlap on root-balled specimens is retained.
- Transplants are at the correct depth relative to the surrounding ground.
- Soil is worked between the roots of bare-root transplants and brought into contact around the rootball of burlapped transplants.
- Mulching and staking when required are carried out as part of the planting operation (see the Mulching and Staking sections on pages 114 and 119).

Fig. 10.1 shows the essential points of good practice when pit planting.

Post-planting checks

Post-planting checks should ensure that:

- The soil is in overall contact with the roots/rootball and there are no large voids between clods.
- Transplants are firmed in.
- When appropriate transplants are locally flood-watered.
- The plants are upright and in the case of trees the main stem (trunk) is vertical.

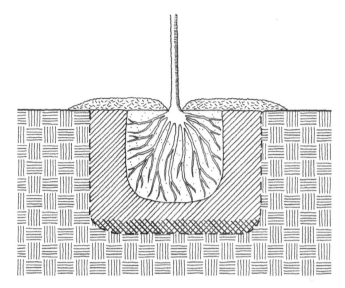

Fig. 10.1. Idealized pit planting. A planting pit must be large enough for back-fill to be placed all round and beneath the rootball and the soil beneath the pit should be broken up. The plant must be upright and at the correct depth.

- Neither the crowns nor the rhizomes of herbaceous plants are planted too high or too low and that woody subjects are at or very slightly below their previous level in the soil profile; ensure that all graft unions are above ground.

Tree and Shrub Planting

The tree shelter

Tree shelters (Fig. 10.2) are transparent plastic tubes developed some 50 years ago by Dr G. Tuley of the Forestry Commission. Research has established that, when combined with a weed-free site, *small* transplants show both a higher survival rate and faster growth when growing *within* tree shelters than the uncovered controls. There is much misunderstanding of the function of tree shelters, as explained in the Forestry Commission's publication *Tree Shelters* (Potter, 1991). Research has found their value to be due to the atmosphere within the shelters having a higher relative humidity and little air movement. The presence of these conditions immediately around the foliage results in stomata remaining open. This leads to a greater intake of carbon dioxide and within a transparent plastic tube an increase in photosynthesis. Tree shelters also protect against predation and aid the efficient use of herbicides.

Tree shelters should be placed over whips immediately after planting, making sure that they are securely staked and that their bases are in contact with the soil all round. They should remain in place for two growing seasons then be removed.

Tree shelters cannot resurrect dead specimens!

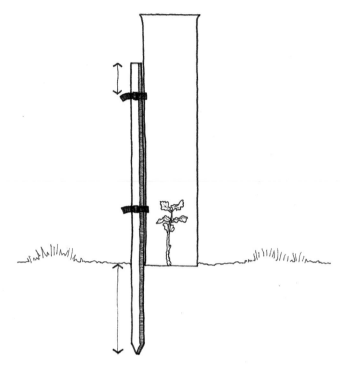

Fig. 10.2. Tree shelters. The tree shelter consists of a transparent plastic tube securely attached to a stake. Note the size of the transplant, the turned top to prevent
chafing, the well-driven stake and the close contact between tube and soil.

Mulching

Perhaps the most useful widespread development in the establishment of plantings
on landscape sites in the last half-century has been the production and use of mulch.
Mulch in this context is a surface layer of 5–7.5 cms of chipped bark or other woody
material. Typically all parts of the canopy of shrubs and trees are satisfactory.

Partly decomposed straw, leaves or herbaceous waste are valuable both
when incorporated into the soil or as a top dressing but should not be used as a
post-planting mulch because they break down too quickly and are then capable of
supporting weed seedlings.

Bark and/or woodchip mulches serve three functions, weed control, moisture
retention and prevention of soil capping. They also provide an attractive finish, particularly where untreated soil surfaces weather down to dust. Such mulches control
weeds by preventing germination from the seed bank in the soil. They do this by
blocking the two stimuli to germination, light and fluctuating diurnal temperature.
They also make a poor seedbed for any weed seeds that arrive from beyond the area;
however, mulches do not kill or even permanently suppress perennial weed ramets.

When tree-derived mulches were first introduced they were based on bark.
There was much discussion on the possible risk of using coniferous material but

neither this nor the use of woodchips is any longer a concern. For more than 20 years a large and prestigious site known to the author has used the total arisings from the chipping of lop and top of both conifer and broadleaved species. This material has proved to be trouble free and extremely effective in controlling weeds and retaining moisture. In this case, the chippings are stored for 2 months to allow the volatiles to volatilize and any white wood fragments to mellow (Fig. 10.3).

Weed mats are made from various materials. They suppress weed ramets, prevent weed-seed germination and conserve soil moisture but, unless they are carefully fitted and secured, weeds can establish at the base of transplants. Weed mats are a far less elegant alternative to mulch. In the example shown in Fig. 10.4 the mat was later covered with woodchip.

Planting sites

Planting on rubble fill and rocky parent material

Excellent results have been achieved with tree and shrub plantings on sites that have insufficient topsoil to provide a 'textbook' soil profile. Typically the root zone

Fig. 10.3. Woodchip mulch. Young shrubs mulched immediately after planting received a top dressing one year later. The site remained weed free throughout this time. Photo courtesy of Mary Payne.

Fig. 10.4. Use of a weed mat. Weed mats are permeable to water but opaque so prevent weed germination and ramet growth of weeds. This example was later covered with a decorative mulch. Photo courtesy of Mikel Pagola.

substrate on such sites consists mainly of rubble from demolished buildings or quantities of broken rock from the local strata. Planting during the dormant season is all but essential. Small transplants (whips) of both trees and shrubs between 600 and 1000 mm in height have been found to be most successful; nevertheless, even with such small specimens the practical task of planting into such material is difficult. It is not easy to make a hole of sufficient depth and width to accommodate a transplant's roots. As with compacted-soil sites, the excavation of pits with a backhoe is preferable to hand-digging. The holes produced by such equipment tend to be larger and the machine shatters any compaction in the surrounding soil. In locations where the excavated material is mostly bricks or large pieces of stone or concrete, it is necessary to bring in sufficient soil to refill the hole.

Even where the substrate is somewhat less hostile, drought before the roots have ramified between the rocks is the biggest risk. This can be reduced by back filling with a 'mixture' of the site material and an organic moisture-retaining 'tree transplanting compost' (green waste) to which a water-retaining gel has been added. The amount of gel depends on site conditions, with 1 kg/ m^3 of compost being a recommended figure. Irrigation during the first two growing seasons is markedly beneficial to both survival and growth rates as shown by Dr Richmond Dutton's extensive research and reported in personal communications.

Planting on sites restricted by hard landscape

In many instances such sites may be thought of as special examples of planting in rubble, as outlined above, and the same procedures should be followed whenever possible. A restricted site does not change the requirements of the transplant.

The excavation of planting sites in hard landscape such as pedestrian areas is likely to expose a hardcore sub-base both below and to the sides of the pit. These restrictions will probably slow root system development, thereby resulting in drought stress during an extended establishment phase. This can be reduced by planting in the dormant period.

Once established, the root systems of street trees successfully ramify between rubble, typically just below the hardtop. This behaviour is often revealed by asphalt or paviours being lifted. To allow for a useful amount of good-quality backfill, planting pits on such sites should be at least 750 × 750 × 750 mm, or five times the volume of the rootball. The base of the pit must be broken up to 250 mm below the depth of the rootball.

To achieve a good contact with the roots, the backfill should be worked around the rootball or, in the case of bare-root transplants, between the remaining roots. In some cases it is not possible to achieve this with freshly excavated material. If on-site topsoil is suitable it should be used; failing this, composted organic matter may be mixed with the excavated material. If organic material has simply been thrown into the base of the pit the specimen should be lifted and replanted.

Planting within areas of newly constructed hard landscape, such as the narrow beds between parking bays in carparks, can present specific difficulties. If the hard surface sub-base has been laid and consolidated overall – that is, including the area beneath the proposed planting beds – there is little hope for the transplants. Roots will be unable to spread and the usual 'dusting' of topsoil will be completely inadequate.

In some cases underground obstructions prevent the size of excavation noted above. In these extreme cases it may be worth resorting to the method recommended back in 1896 by Stringfellow in *The New Horticulture*. In summary, this would require selecting appropriate trees such as *Sorbus* spp. and *Tilia* spp. and planting in the dormant season using bare-root, light-standard transplants that had been given extensive root pruning and crown reduction. The transplants will need supporting both to prevent them from leaning out of true and to hold them steady until a new root system develops.

On restricted sites and those where backfill is little better than rubble, specimens should be 'flood-watered' immediately following planting. This saturates the backfill and settles it into close contact with the roots. Irrigation will also be required during the first growing season (see the Irrigation of tree transplants section below).

In some extreme conditions the planting pit may have to be large enough to accommodate the root system throughout the life of the specimen. Assuming a small tree species has been selected then a 1.5 m × 1.5 m × 1 m pit should be excavated. The nature of the backfill will depend on local factors. A specification for a substrate that is both load bearing for light vehicles and root-penetrable is available; essentially it is a mixture of very coarse aggregate and high-grade topsoil/potting compost. This specification is based on Danish research by Dr Thomas Randrup. For an English explanation, see the National Urban Forestry Unit (2000) *Urban Forestry in Practice: Load Bearing Soils for Trees in Paved Areas.*

Planting sites surrounded by a water-impenetrable hard surface must have a porous surface of at least 1 m^2 immediately around the trunk to allow air (oxygen) and surface water from both rainfall and irrigation to reach the root zone. To prevent the exposed soil from compressing and capping it should be protected with an open or perforated grid. The evidence that many mature street trees do not have such a porous area should not be taken as a reason for omitting this biologically important design detail. In many cases the success of such trees can be attributed to older, less-engineered forms of hard-landscape construction.

The setting of the soil surface relative to the surrounding area must be judged on a site-by-site basis. The advantage of finishing the soil surface below its surrounds is that both surface runoff and irrigation water can be directed to the transplant's root system. The danger comes in locations where the runoff may be polluted from road washings and/or salt used to de-ice sidewalks. The reciprocal condition where a tree's roots may become a problem by either creating uneven surfaces or by being the primary cause of damage to buildings may determine planting locations. This complex topic is explored by Cutler and Richardson (1989) in *Tree Roots and Buildings.*

Planting in arable land

The initial stages of this work are, in essence, the same for both trees and shrubs. In each case the planting holes (pits) must be appropriate for the size of the rootball. The soil to the sides of the hole should be in a friable state following cultivation. Agricultural tillage equipment is unlikely to penetrate for more than 150 mm to 200 mm so unless the area has been ripped, as described on page 23, Easing compaction section, the soil beneath the pits dug for larger specimens may still be compacted. In such cases it must be broken up. Irrigation is typically a simple matter of applying water to the base of each plant. When driving stakes into cultivated beds it is important to ensure that they reach the undisturbed horizons because the cultivated topsoil will be too loose to hold them. Weed colonization can be a severe problem. Mulch has been shown to be an excellent way of preventing this. In many cases, it is simpler and more cost-effective to apply it over the whole of the planted area (see the Mulching section above).

Planting on undisturbed grassland

Such sites should not be cultivated overall but allowed to retain their soil profile. Potentially competitive close-mown and rough herbage must be killed off. Depending on the nature and density of the proposed planting, this is best achieved either by an overall application of herbicide or by spot treatments restricted to the planting locations. In the latter case the herbage should be killed for at least a 0.5 m radius around each planting site, to produce a 1 m diameter weed-free circle.

Tillage is typically restricted to the digging of planting pits. Their construction, and the appropriate methods of planting, irrigation and mulching around the transplants, are as noted above. In the case of whips planted in lines to form privacy screens or rural hedges, a 1 m-wide strip is killed off. Where site conditions allow, tillage should take the form of ripping down the planting lines to a depth of 400 mm with a tractor-mounted agricultural subsoiler. This shatters any compaction and makes 'notch planting' using a forester's planting spade easier. It usually significantly improves both 'take' and subsequent growth rates.

Staking standard tree transplants

Standard tree transplants of all sizes and in most locations are likely to require staking for between 3 and 5 years using one of the methods illustrated in Fig. 10.5. The stems/trunks of good-quality transplants should not need support; the function of the stake(s) is to provide stability below ground to allow roots to spread outwards while at the same time keeping the trunk upright. The investment in using one of the well-tested methods shown in the illustration should be linked to site exposure; in extreme locations the use of smaller specimens is usually preferable to large standards.

Two common and indelible problems that have arisen with the planting of larger standard trees can be traced to poor staking. One is where tree-trunks lean at a drunken angle beneath upright canopies. This is usually due to the omission of stakes, or stakes not secure in the ground, or the ties between stake and tree breaking. The other is damage to tree-trunks or the lowest branches caused by their rubbing on the top of stakes or crossbars. Such chafing results from poor security or lack of packing between tree and support.

There are several satisfactory and well-documented methods of staking; see, for example, Rushforth (1987) in *The Hillier Book of Tree Planting and Management*. Whenever high stakes are used, a minimum of two ties is required. Trees with well-developed rootballs may be secured with below-ground anchors, thereby avoiding stakes or guys becoming obstructions.

In recent years much has been written about the benefit of using short 1–1.5 m stakes to support traditional standards. The argument is that in the case of well-grown transplants there is a need to stabilize the roots or rootball but that the tree's trunk and canopy should not need support. It is known that stem and foliage movement (thigmomorphogenesis) tends to result in an increase in stem diameter so producing a sturdier specimen. In practice the choice between low and high staking is often situation based, with high stakes helping to reduce damage by vandals.

Irrigation of tree transplants

Watering throughout the first growing season after planting is beneficial, and often essential in most years and many situations; frequently the benefit extends to the

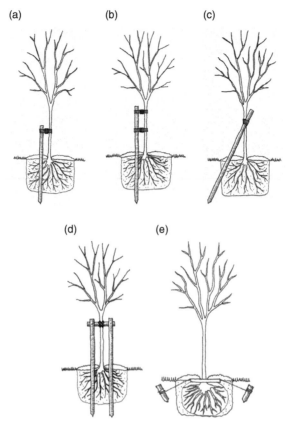

Fig. 10.5. Tree staking. The object of tree staking is to hold the tree upright and steady whilst it re-establishes its root system, hence the stake must be securely driven. All above-ground staking methods should use buffer pads to prevent damage to the tree. (a) Short stake, suitable for bush and standard trees of a size below that of heavy standard. (b) Traditional long stake, suitable for standard trees of a size below that of extra-heavy standard. Such stakes may be fitted with guy-ropes for extra stability if the location is suitable. (c) Angled stake, similar to the short stake but of particular value in windy locations. Note the stake should point into the prevailing wind. (d) Twin stakes with cross bar. A secure support for the larger grades of tree transplant. Note the cross-bar must be buffered to prevent damage to the tree stem. (e) Ground anchors (below-ground guying). A way of securing transplants with large rootballs in locations where staking is considered inappropriate.

second year. For large tree transplants the challenge is to ensure that the water gets to roots within the rootball and its immediate surrounds. The benefits of irrigation have been stressed by the researchers based at the University of Liverpool; see, for example, Bradshaw, Hunt and Walmsley (1995) *Trees in the Urban Landscape*. The form of irrigation may be as basic as flooding the planting site, whereas,

at the other extreme, some very prestigious systems used on semi-mature transplants have built-in automatic watering systems; these are typically specified and
installed by specialists. In two simple but effective approaches, the water is delivered to the root zone via a vertical pipe into which a hosepipe can be pushed. In
one design the transplant's rootball is set on a 150 mm bed of coarse gravel upon
which the vertical pipe rests. In the alternative design (Fig. 10.6), the tube is made
of a length of perforated drainage pipe coiled around the base of the trunk on top
of the rootball. In both cases water application should continue until the water
overflows from the feeder pipe.

Pruning directly associated with transplanting

There should be no need to prune nursery-grown trees and shrubs to aid their
establishment. Trees and large shrubs that are to be moved within the site are,
however, likely to benefit from a reduction in their canopy, either by thinning out
the number of branches (stems) or by reducing their height and spread. Coppicing
has proved to be a successful if drastic aid to rescuing precious specimens.

Hedge transplants, typically of native species, are often cut back to around
150 mm above ground level, either at the time of planting or after one growing
season (see page 132, Cutting back and coppicing section). This removes

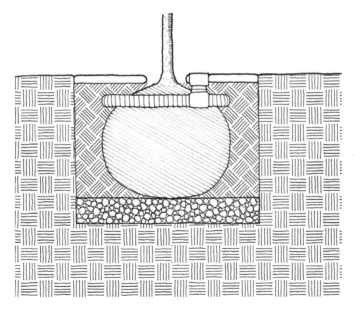

Fig. 10.6. Irrigation of large transplants. This simple elaboration of hose watering
ensures that water goes directly to the transplant's root system and allows the required volume of water to be discharged in a short time. Note the hose connection
point, the coiled hose resting on the rootball and the aggregate filled 'reservoir'.

apical dominance and so induces the production of several stems that form the basis of a thick, stock-proof hedge.

Shrub planting

The same good practices of rootball inspection, watering, determination of planting-hole size and planting depth, firming of backfill and mulching apply just as much to shrubs as to trees. The form of the root mass varies between species; particular care is needed when planting very fine-rooted types such as *Erica* and *Rhododendron*. Such plants survive well within their rootball but are slow to root out into the surrounding soil and hence risk drying out. As emphasized throughout this book, the backfill must have a fine crumb structure or particle grade so that when firmed down it makes good contact with the rootball (see also page 129, the Irrigation at and after planting section).

Planting and supporting climbers

The most common mistake is to mismatch the plant's method of climbing with the proposed support (see page 89, Climbers section). Climbers are almost always transplanted from containers. They should be planted as close to their support as possible. Where such obstacles as wall footings or the bole of a tree prevent close planting, the stem of the climber should be brought to the support as near to the ground as possible. All climbers, including the self-clinging types, need assistance to link them to their permanent supports, which is usually provided by tying them to carefully positioned canes.

Planting Herbaceous Perennials

Site preparation for plants which die down to their roots every winter requires somewhat more care than that needed by plants with woody stems. Preparation is based on overall cultivation to a depth of a minimum of 250 mm, finishing to running levels and with a reasonable tilth. To grow vigorously, many herbaceous perennials require a reasonably fertile soil; hence on many sites the pre-planting cultivation should include the incorporation of organic matter and a dressing of general fertilizer. A common formulation is 20% nitrogen, 10% phosphorus, 10% potassium sold as 20:10:10 NPK and applied at 20 g/m^2. This cultivation also provides the opportunity to remove perennial weed ramets – a vital consideration because it is very difficult to eradicate them later within such plantings either by hoeing or by the safe use of herbicides.

It is important to plant at the correct depth so that the apical bud of rosette-forming species or the surface rhizomes of plants such as *Bergenia* and the bearded iris are at ground level, neither buried nor left supported by exposed roots.

Spacing and layout

Mass planting to achieve ground cover

The massed planting of herbaceous perennials to achieve ground cover is most effectively achieved by using small, maximum 2 litre, young, actively growing transplants; these should be planted so as to achieve complete ground cover within two growing seasons. For example, to achieve this *Bergenia* cultivars require planting at five to the square metre.

Recently introduced forms of herbaceous plantings

As noted in Chapter 7, herbaceous plantings lend themselves to 'out-of-garden' landscapes. They are not sufficiently commonplace as yet, however, to include in a planting contract without specialist guidance.

Over and above the requirements common to all herbaceous plantings noted above the following appear to be of particular importance:

- The size of the area of each taxon and its repetition;
- The combinations of taxa to produce plant communities;
- The spacing, density, form and size of the transplants of each taxon in anticipation of some natural spread;
- The introduction of transitory eye-catching emergents and their anticipated self-seeding (often biennials).

Where the first dormant season falls within the planter's responsibility, the approach to, and timing of, the removal of dead biomass is critical. The autumn and early winter retention of dry stems, leaves and flower heads may be intended.

Bulb Planting

Bulbs of spring-flowering species are available from September until November. Bulbs should be planted at approximately twice their own depth. The degree of precision required regarding bulb orientation has been investigated on behalf of commercial growers. It was found that there was no significant difference in the performance of those planted on their sides rather than upright.

On bare land intended to become turf it is far easier to plant before turf laying or seeding. Large-scale planting into an established sward is best done by temporarily lifting the turf and cultivating the exposed soil to a depth sufficient to allow the bulbs to be set at the required depth before relaying the turf. Lifting the turf gives the opportunity to create an informal distribution of bulbs and avoid planting in straight lines. Many designers of 'bulbs in grass' schemes favour planting in 'clumps'. As shown in Chapter 7, Fig. 7.17, clumps of between 10 and 15 bulbs depending on the species should be most closely planted at their centres, suggesting the spread from a single individual bulb. Using narcissus, an average of 30 to 50 bulbs per square metre gives satisfactory results. When planting small bulbs into turf, an electric drill fitted with a wood auger has proved both effective and fast.

As part of a spring bedding scheme, tulips and hyacinths are the most popular – planted either as a single display or in combination with other spring-flowering bedding plants; typical densities for tulips alone range from 80 to 100 per square metre. In combination with annual bedding plants this drops to between 30 and 50.

Bedding Plant Planting

Bedding plants are most often encountered in public open space providing colour in those containers often referred to as 'street furniture'. There is a very wide range of 'off the peg' containers but whichever is selected must provide for adequate drainage to prevent waterlogging. Filling such containers with un-ameliorated topsoil is generally unsatisfactory. A 'potting compost' using the formula developed by The John Innes Institute and described by Lawrence and Newell (1939) in *Seed and Potting Composts* is ideal.

The more traditional 'flower beds cut into turf' remain popular providers of seasonal floral colour. As with the isolated containers, the nature and condition of the soil is critical if the transplants are to establish quickly enough to grow away and produce an impressive display through the season. Within the root zone the soil should be ameliorated to bring it as near to a potting compost in composition and structure as possible (see page 28, Soil additives section).

While there may be an understandable desire to produce an instant effect, both the right planting date and condition of transplant are essential to achieve season-long success. Transplants should have no more than one flower open; more floriferous specimens are unlikely to make sufficient new growth to give a long display season. In temperate Europe, autumn planting should be done between the start of October and mid-November, and spring planting between mid-May and mid-June. To achieve a satisfactory display it is essential to plant at the correct density. Although this varies from species to species, a minimum number of 20 plants per square metre should be allowed for.

Turf Laying

The preparation of the site must produce an evenly consolidated, level – or 'to falls' – surface so that when laid the turf will be at the required height with regard to borders, paths and curbs. (It should be noted here that turves are now cut thinner than in the past.) Adjusting levels afterwards very rarely gives satisfactory results.

Turf quality assessment should include sward density and evenness. The turf should not be allowed to dry out or remain long in the rolls where, depending on the season, it will overheat and/or blanch. Turf is traditionally laid 'as brickwork' – that is with no straight joints between the turves at least in one direction; however turf is now delivered to the professional user in long and wide rolls which greatly reduces the number of joints. Once laid, areas requiring a lawn-quality finish should have their joints sanded-in; in most years it will be necessary to irrigate until the turf has rooted into the soil, particularly when using thin turves.

Direct Seeding on Site

The tasks involved in seed sowing, although concerning a wide range of plants and aimed at achieving diverse design features, nevertheless have several aspects in common, including site preparation, seed sowing and the immediate post-sowing husbandry.

Following cultivation the seedbed should be lightly consolidated before the surface is tilled to produce an even tilth of a fineness appropriate for the size of the seed. Uneven seedbeds result in uneven moisture at the soil surface and hence erratic germination. The seed should be sown evenly at the desired density and depth. After sowing, a light roll may be used to ensure the seed makes contact with the substrate and thereby with moisture. Details of each kind of direct seeding are available in the literature and from specialists noted below under the appropriate headings.

Direct seeding of amenity turf

Grass seed will germinate and establish seedlings on a very wide range of soils, but the success of the resulting turf sward is as dependent on the preparatory work prior to sowing as it is on the site's future management.

Many areas designated as turf depend on imported topsoil. Before the topsoil is spread it is essential that the *in situ* substrate is loose enough to ensure free drainage and root penetration. Freedom from compaction is of greater importance than the nature of the substrate. Given friable subsoil conditions, turf for general-amenity use needs little more topsoil than that required to form a good seedbed. In most cases that is around 150 mm but with exceptionally good subsoil it can be as little as 100 mm.

When soil is spread between the area's running levels, particular attention should be given to the relationship between finished soil and path or curb edges, with allowance for settlement. The selection of an appropriate seed mix should be based on the role the turf has in the design. This ranges from meadowland to roadside verge, from public amenity/recreational site and sportsground to fine lawn. With all but highly fertile soils, grass seedlings produce a stronger sward following a dressing of specifically formulated pre-seeding fertilizer. The rate per unit area will depend on the concentration of the nutrients in the mixture. The typical recommended range is between 30 and 50g/m^2. The seedbed requires a reasonable tilth without reducing the soil crumbs to dust as many amateur gardeners do.

Seeding is typically done by broadcasting by either hand or machine, or by hydroseeding. It can take place, weather permitting, at any time between March and October in lowland Britain; many grasses germinate at 7°C. It is preferable to avoid sowing in high summer when there is a risk of drought. Under such conditions irrigation will be needed during the germination period.

Seed size varies greatly between turf grass species, with numbers per gram ranging between 500 and 3000. Clearly this large range will influence the seed rate, as does the nature of the site and its intended use – all variables that negate a precise single recommendation. As a general indication, the sowing rate for a typical mix should not exceed 25 g/m². Having excessive numbers of seedlings encourages fungal pathogens and early competition between seedlings.

Post-germination, the appearance of weed seedlings often causes concern but the vast majority of these are killed out by the first mowing or 'topping' a few weeks after the grass seedlings are well established.

Direct tree and shrub seeding

When direct woody plant seeding is considered, it is advisable to contact specialist contractors at the outset because seed has to be sourced and possibly stored (or not, depending on the species selected). Such contractors usually offer a full service, which will include the following advice:

- Appropriate species to achieve the design intent; reputable suppliers also provide details of seed origin;
- Best time to sow;
- Appropriate sowing technique(s), which may be broadcasting or drilling or hydroseeding. Tree seeds vary in size from, for example, horse chestnut (*Aesculus hippocastanum*) to the birches (*Betula* spp.). Differing size, viability and germination biology may combine to require mixtures to be sown at different times and depths, and even using different methods;
- Sowing density.

Such specialist contractors should then carry out the following operations in this order:

- Site preparation, which should include a check on the depth of the root-penetrable substrate and any remedial tillage that is required to achieve this and the elimination of perennial weeds;
- Establishment of a seedbed with a surface tilth of soil crumbs ranging from 10 mm to 2 mm, depending on the seed size of the selected species.

To aid even distribution fine seeds should be mixed with a spreader, which may range from sand to sawdust. Depending on their size, the seed may be very lightly raked in, followed by a light firming to ensure the seed makes contact with the substrate and thereby with moisture. The forest nursery practice of covering a seedbed with a 100 mm layer of coarse sand or grit may be considered. It is often beneficial to cover the seedbed with a thin layer of straw to keep the surface from drying out too much during the period between seed sowing and germination, at which point it must be removed. The germination and seedling growth of large-seeded species benefits from a seedbed covering of chopped straw, which is allowed to remain to become a mulch.

Post-sowing site management may have to include irrigation and also measures to exclude predators and control weeds (tree seedlings are severely affected by weed competition). To achieve the optimum plant stand, either thinning out or gapping up (beating up) may be needed between the first and second growing seasons.

Direct seeding of native and exotic annuals

Unlike most planting practices, this has undergone significant research and refinement. It is therefore advisable to deal with reputable suppliers, who will provide advice on seed mixes and rates, since to deliver the design intent these should be decided on a site-by-site basis.

Site preparation is the same for native and exotic annuals. The usual aim is to achieve a cultivation depth of 200 mm above a free-draining subsoil. The surface tilth should be appropriate to the small size of the seed of most annuals, but this need not exclude rock or rubble fragments as found on brownfield sites; soil nutrient levels should be low because high fertility tends to favour rank growth and weeds such as docks and thistles.

Native annuals may be sown in either autumn or spring, whereas the exotic species are best held back until late spring because some types may be frost sensitive. In order to get an even distribution and the required low seeding density the seed is often mixed with a spreader of sand or fine sawdust. Seed is broadcast by hand and, depending on the fineness of the tilth, very lightly raked in but not deeply buried. It is important to keep the soil surface moist during the germination period. As with tree seeds on exposed sites or in dry conditions, it has been found beneficial to cover the seedbed with a thin layer of straw for the period between sowing and germination. Hydroseeding has proved to be successful; when used, the carrier acts as a mulch.

Direct seeding of native perennial herbs (species-rich meadows)

Establishing the kinds of wild flowers found in a species-rich meadow may seem easy but it has proved otherwise. Overcoming the problems noted on page 103, in the section on Establishing native perennial herbs, has resulted in a number of critical recommendations based on well-managed research. Recommendations vary, depending on the specific plant community intended and the site conditions; it is very advisable to follow them as closely as possible.

The following guidelines are intended only as basic, general ones. Specialist instruction is strongly advised. Soil preparation and seed sowing is as noted above for annuals but typically with a lower seed density. Seed germination may be erratic and occur over a long period. This, together with the vigour and invasive nature or otherwise of each species, should be taken into account when determining its contribution to the mix. There is likely to be little in the way of flower in the first year after sowing and some annuals may be included to augment the poor display. Grass must be regarded as a weed during the time that the forbs establish. Many experts recommend omitting grass from the seed mix. A low soil fertility is desirable because it will favour the forbs at the expense of the grass.

Care in the Establishment Phase

11

Irrigation at and after Planting

The provision of sufficient water is one of the two most important horticultural activities throughout the early care of a new planting, the other being weed control. At the time of planting both container-formed and burlap-wrapped rootballs should be moist. Outside the dormant season, the soil round each newly planted rootball should be flood-watered immediately after planting.

In many years irrigation is required during the defects liability period. Recently planted stock with its reduced or constrained roots suffers from drought stress much sooner than established specimens. This is particularly so if there is poor contact between transplant and soil – hence the need to 'water in'. Transplants can draw moisture only from the soil immediately around the spread of retained roots, a zone far smaller than that round an undisturbed specimen. Container-grown stock will have been regularly watered such that their small soil volume never or rarely dried out. After planting they will continue to depend on the moisture within the rootball. It is relatively easy to wet the surrounding soil; getting a dry rootball to take up water is difficult.

Once established, few sensibly designed landscape areas will need irrigation but a satisfactory method of irrigation must be in place to cover the establishment phase, which may extend beyond the defects liability period. If responsibility for the soft landscape is handed over to a maintenance contractor it is prudent to negotiate for irrigation visits to be made as required in times of drought rather than a fixed number regardless of conditions.

Hose watering is the simplest form of short-term root-zone irrigation; however, there are several commercial designs for permanent irrigation systems. On large plantings, it is possible to install drip or seep systems with a low-volume discharge. Once these are no longer required they should be capped off. Tree transplant irrigation is discussed on page 119.

It is possible to tackle drought stress by addressing water loss. Moisture retention varies with soil texture but is generally greatly reduced by weeds. It can, however, be much improved by applying an appropriate mulch as described on page 114, Mulching section).

Reducing moisture loss from the stems and leaves of transplants has been investigated and both windbreaks and, in particular, the tree shelters described on page 113 have been shown to be effective.

Anti-desiccation sprays or dips based on emulsified polymers are marketed to reduce water loss from foliage. In practice, however, when applied either immediately before or after planting they have been found to be of limited value and are rarely used on landscape sites in Britain.

Detection and Prevention of Damage to New Plantings

Protection of soil and plants from ongoing building work

If horticultural work must be carried out while building activities are still ongoing, both soiled areas and plantings are vulnerable to damage by vehicles and human actions such as dumping, storage and disposal of washings. Notices and fencing at an early stage after soiling is advisable but developing a good working relationship with the main contractor is probably the most cost-effective approach.

Detection and prevention of damage by mammals and birds

Rabbits

On many sites rabbits are by far the most serious pest. In rare instances, it is possible to eradicate the local population before work starts but more often the only feasible action is to protect transplants immediately after planting by fencing either the whole planting or individual specimens.

There are three commonly used forms of protection:

- Wire netting is usually employed to surround whole sites or plantings. The standard recommendation is for 1.2 m high galvanized wire netting with 31 mm hexagonal mesh. It is essential to bury the base of the netting 150 mm deep.
- Plastic spills or spirals to protect individual trees. Their use to protect hedging transplants may, however, suppress laterals and result in a hedge with few stems below 0.5 m.
- Tree shelters confer the incidental benefit of protection from predators, as well as from herbicide spray and strimmer damage (see page 113).

To gain these advantages the tubes must be securely staked and pressed well into the soil surface.

Details of the use of each form of protection are given in the trade literature.

Smaller rodents

Mice, or more often voles, can cause very serious losses in winter among newly planted trees, shrubs and bulbs. Typically with woody plants the damage is in the form of bark ringing at ground level. Mulch provides the animals with ideal cover; it should be pulled back from transplant stems in late autumn. As with rabbits, plastic spills or spirals can protect individual trees if their bases are firmly pushed into the soil. Chemical repellents are offered but are found to give limited protection.

Deer

Deer of several species make up a third mammalian group whose increased numbers have caused serious damage in recent years. They have been known to eat a wide range of herbaceous plants but are more notorious for browsing in the form of either bark-stripping the woody stems of trees and shrubs or eating the young shoots and foliage. As with rodent control, chemical repellents are offered but again are found to give limited protection. Deer fencing is seldom feasible on development sites but in rural locations this expensive solution may be appropriate or indeed essential. Trout and Pepper (2006) *Forest Fencing*, a Forestry Commission Technical Guide, provide details for both rabbit and deer fencing.

Birds

The two most common bird-derived problems associated with recent horticultural work are the scattering of mulch and damage to recently sown areas. Mulch displacement is usually due to the birds searching for insects. This suggests that the mulch is providing a habitat for insects, which means that the material is too old and/or fine-textured to be suitable for weed suppression (see page 114, Mulching section).

Damage to seeded areas, including lawns, may be caused by birds eating the seed and/or dusting in the surface tilth. Bird scaring is hardly ever practicable on development sites; however, seeded areas can be netted for the period between sowing and germination and bird-repellent-dressed grass seed is available.

People and farm animals

Protection from people may be at odds with the concept of amenity and today we rarely see notices ordering us to 'keep off the grass' but newly sown turf and planted areas need protecting by a sign, albeit more polite, and/or by barrier tape or the old-established but still effective use of bamboo hoops.

Accidental damage is also caused by the very people employed to maintain the site. To be fair, it is not they but their strimmers that severely damage and in many cases kill young transplants. Strimmer guards fitted around the base of the trees will prevent this, although in many cases a weed- / grass-free area around the tree is a preferable solution.

Rural amenity sites, including those established for nature conservation, often have grazing farm animals as an essential part of their management. Areas of shrub and tree planting require fencing off for several years until the plants are robust enough to withstand the animals' attention.

Specimen trees require protection by one of the many designs of wooden 'tree guard' that have replaced the forged-iron Victorian models. It is essential that their height and width takes account of both the reach of farm stock and the developing crown of the tree. It is not uncommon to find that the guard erected to protect a tree is damaging its trunk and branches by tree and guard rubbing together.

Pruning

Good-quality stock will require no pruning before planting, any tidying up having been done in the nursery before dispatch. Material lifted from within the site may need thinning or canopy reduction, as pointed out on page 36, Translocation of trees and shrubs section.

Pruning during the establishment period

After one year, so within the defects liability period, standard tree transplants may produce both lateral 'water shoots' and start to develop double leaders; all water shoots and the weaker of the two leaders should be removed.

The unripened tips of shoots on transplants sometimes die back and can be responsible for disease entry and moisture loss from more proximal parts of the plant. Their removal in the early part of the first growing season also encourages growth from lateral buds. In most plantings, the only pruning required at the end of the first growing season, other than that noted below, will be on specific subjects such as roses and lavender.

Cutting back and coppicing

Whips planted to form hedges may be cut hard back to achieve shoots low down on their stems. The cutting back of seedling transplants of some tree species, particularly oak, after one growing season has been found to produce strong juvenile regrowth, markedly outstripping that on those left unpruned. Shrub transplants to be coppiced for woodland understory or winter bark colour are usually left to establish for 2 years before cutting back. In many cases the most important instruction during the establishment period is to *not* start the annual cycle of indiscriminate clipping with mechanical hedge trimmers that ruins so many designs.

Weed Control

Pre-planting weed control is discussed on page 107, Pre-tillage weed control section.

Poor weed control in the establishment phase has been proved to have a great effect on the growth of woody transplants, as proved by a most revealing experiment conducted by the Forestry Commission at Alice Holt from 1984 to 1986; when the impact of weeds on tree transplants was made manifestly clear (Fig. 11.1). For more details, see Davies (1987) *Trees and Weeds* in the Forestry Commission Handbook 2.

It is evident that a bare, grass-free area of a minimum of 1.2 m diameter should be maintained for 3 years round trees planted into turf. In other locations, the application of mulch should reduce the weed problem to a few survivors and some windborne introductions, which often include weed tree seedlings.

Post-planting weed control should be based on herbicide spot treatments or hand-pulling before any weeds flower. Hoeing requires care, skill and close supervision to avoid damage to recent plantings; digging between transplants is totally inappropriate. Allowing weeds to seed increases the size of the seed bank, thereby putting problems in store for future years. Where necessary the mulch layer must be topped up to its original depth.

Fig. 11.1. Weed competition trial. This Forestry Commission investigation into the effect of grass/weed competition on small tree transplants produced dramatic results that are a valuable lesson in site management. Image Crown Copyright courtesy of Forest Research.

Provision of a Management Programme

Throughout this book, emphasis has been placed on achieving the design intent by *a combination of the specific nature of the plants and their specific nurture*. Thus, the selection of the plants is only half the process. To achieve the desired result the designer should indicate the style or form of the management of the plantings. Where appropriate such guidance should include instructions for the timing of specific activities critical to the intended development of the site, including the thinning and coppicing of woodland areas, and the pruning or non-pruning of woody plantings, the required training of isolated specimens and the removal of stakes and tree guards from individual trees.

If commissioned, a more detailed management schedule should include an outline routine maintenance programme, again directly linked to the design intent; this might include, for example, the identification of appropriate mowing regimes for turfed areas, be they prestige lawn or species-rich grassland, and/or the specific treatments required by different styles of herbaceous plantings (e.g. see Hitchmough (1994) *Urban Landscape Management*).

Inevitably this book has focused on problems and difficulties but it should bring some cheer to recognize that only a small number of the more challenging situations and problems discussed are likely to be experienced during a single project.

References

References in the Text

Black, C.A. (1968) *Soil–Plant Relationships,* 2nd edn. Wiley, London.

Bradshaw, A.D., Hunt, B. and Walmsley, T. (1995) *Trees in the Urban Landscape*. Spons, London.

Clamp, H. (1995) *Spons Landscape Contract Handbook,* 2nd edn. Spons, London.

Cobham, R. (ed.) (1990) *Amenity Landscape Management*. E. & F.N. Spons, London.

Cutler, D.F. and Richardson, I.B.K. (1989) *Tree Roots and Buildings*. Construction Press, London.

Cubey, J. (ed.) (2014–15) RHS Plant Finder. Dorling Kindersley, London.

Darthuizer Vademecum (2005) Darhuizer Boomwekerijen BV, Holland.

Davies, R.J. (1987) Trees and weeds. In: *Weed Control for Successful Tree Establishment Handbook 2*. Forestry Commission, Alice Holt, Surrey.

Department of Communities and Local Government (2014) *Tree Preservation Orders and Trees in Conservation Areas*. HMSO, London.

Department of Transport (1992) *The Good Roads Guide. Environmental Design Guide for Inter-Urban Roads*. HMSO, London.

Dunnett, N. and Hitchmough, J. (eds) (2004) *The Dynamic Landscape*. Spons Press, London.

Dutton, R.A. (1991) An analysis of the critical stages in urban tree establishment. Doctoral dissertation, University of Liverpool, Liverpool, UK.

Dutton, R.A. and Bradshaw, A.D. (1982) Land Reclamation in Cities. Department of the Environment, HMSO, London.

Evison, J.R.B. (1958) *Gardening for Display*. Collingridge, London.

Field, C. (ed.) (1991) *A Seed in Time*. The Second UK/International Conference on Urban Forestry. Wolverhampton Polytechnic, UK.

Garner, R.J. (1947) *The Grafter's Handbook*. Cassell, London.

Gregory, P.J. and Northcliff, S. (2013) *Soil Conditions and Plant Growth*. Wiley–Blackwell, London.

Grime, J.P. (1979) *Plant Strategies and Vegetation Processes*. John Wiley, London.

Hansen, R. and Stahl, F. (1993) *Perennials and their Garden Habitats*, 4th edn. Cambridge University Press, Cambridge, UK.

Hillier Manual of Trees and Shrubs (2014) Royal Horticultural Society, London.

Hitchmough, J.D. (1994) *Urban Landscape Management*. Inkata, Sydney.

Hitchmough, J. and Fieldhouse, K. (eds) (2004) *Plant User Handbook*. Blackwell, Oxford, UK.

Hartmann, H.T. and Kester, D.E. (2002) *Plant Propagation: Principles and Practices*. Prentice Hall, Oxford, UK.

HTA (2002) *National Plant Specification*. Horticultural Trades Association, Theale, Reading, Berkshire.

Joint Liaison Committee on Plant Supplies (1981) *Herbaceous Plants: Exotic and British Native*. Joint Liaison Committee on Plant Supplies, UK.

Joint Liaison Committee on Plant Supplies (1981) *Code of practice for Plant Handling*. Joint Liaison Committee on Plant Supplies. UK.

Jones, H.G. (2014) *Plants and Microclimate*, 3rd edn. Cambridge University Press, Cambridge, UK.

Kelway, C. (1962) *Seaside Gardening*. Collingridge, London

Kendle, A.D. (1996) The nature of soils on landscape sites and their effect on plants. In: Thoday, P. and Wilson, J. (eds) *Landscape Plants*. Cheltenham & Gloucester College of Higher Education, Cheltenham, UK.

Kendle, A.D. and Forbes, S. (1997) *Urban Conservation*, E. & F.N. Spon, London.

King, L.J. (1966) *Weeds of the World*. Leonard Hill, London.

Lawrence, W.J.C. and Newell, J. (1939) *Seed and Potting Composts*. Allen & Unwin, London.

Le Sueur, A.D.C. (1951) *Hedges, Shelter-Belts and Screen*. Country Life, London.

Luscombe, G. and Scott, R. (2004) *Wild flowers Work*. Landlife, Liverpool, UK.

Luscombe, G., Scott, R. and Young, D. (2008) *Soil Inversion Works*. Landlife, Liverpool, UK.

Medhurst, J. (2013) *Trees in London*. Available at: www.johnmedhurstlandscape.co.uk/trees/ (Accessed 16 April 2016).

National Planning Policy Framework (2014) Tree preservation orders and trees in conservation areas. Department of Communities and Local Government, UK. Available at: http://planningguidance.communities.gov.uk/blog/guidance/tree-preservation-orders/ (Accessed 14 April 2016).

National Urban Forestry Unit (2000) *Urban Forestry in Practice: Load Bearing Soils for Trees in Paved Areas*. National Urban Forestry Unit, Wolverhampton, UK.

Naylor, R. (2002) *Weed Management Handbook*, 9th edn. Blackwell, Oxford, UK.

Pollock, M. (1984) *Shelter Hedges and Trees*, 4th edn. HMSO, London.

Potter, M.J. (1991) *Treeshelters Forestry Commission Handbook 7*, HMSO, UK.

Rehder, A. (1940) *Manual of Cultivated Trees and Shrubs Hardy in North America*. MacMillan, New York.

Roads Beautifying Association (1930) *Roadside Planting*. Country Life, London.

Royal Horticultural Society (2015) *Award of Garden Merit Plants*. Available at: www.rhs.org.uk/plants/trials-awards/award-of-garden-merit (Accessed 14 April 2016).

Rushforth, K.D. (1987) *The Hillier Book of Tree Planting and Management*. David & Charles, Devon, UK.

Stringfellow, H.M. (1896) *The New Horticulture*. Galveston, Texas.

Thoday, P. (2007) *Two Blades of Grass*. The Story of Cultivation. Thoday Associates, Box Hill, Wiltshire, UK.

Thoday, P. (2013) *Cultivar: The Story of Man-Made Plants*. Thoday Associates, Box Hill, Wiltshire, UK.

Thoday, P. and Wilson, J. (eds) (1996) *Landscape Plants*. Cheltenham & Gloucester College of Higher Education, Cheltenham, UK.

Trout, R.C. and Pepper, H.W. (2006) *Forest Fencing*. Forestry Commission Technical Guide, Edinburgh.

Wandell, W.N. (1989) *Handbook of Landscape Tree Cultivars*. East Prairie Publishing Company, Gladstone, Illinois.

White, R.E. (2005) *The Principles and Practices of Soil Science*, 4th edn. Blackwell, Oxford.

Willoughby, I., Stokes, V., Poole, White, J.E.J. and Hodge, S.J. (2007) The potential of 44 native and non native tree species. Forestry Commission, Alice Holt, Surrey.

Winter, E.J. (1974) *Water, Soil and the Plant*. Macmillan, London.

Further Reading

Bean, W.J. (1973–76) *Trees and Shrubs Hardy in the British Isles*. Murray, London.

Bengtsson, R. (2005) Variation in common lime (Tiliax europaea L.) in Swedish Gardens of the 17th and 18th centuries. Doctoral dissertation, Swedish University of Agricultural Sciences, Alnarp, Sweden.

Booth, C. (1957) *An Encyclopaedia of Annual and Biennial Garden Plants*. Faber, London.

Bowler, D.G. (1980) *The Drainage of Wet Soils*. Hodder & Stoughton, London.

Christians, N.E. (2011) *Fundamentals of Turf Grass Management*. John Wiley and Sons Ltd, UK.

Davies, B., Eagle, D. and Finney, B. (1972) *Soil Management*. Farming Press, Ipswich, UK.

Emmons, R.D. and Rossi, F. (2015) *Turfgrass Science and Management*. Cengage Learning Inc. Boston, Massachusetts.

Evelyn, J. (1664) *Sylva, or A Discourse of Forest-Trees, and the Propagation of Timber in His Majesties Dominions*. London.

Forestry Commission (2002) *Forest Reproductive Material*. Great Britain.

Gilbert, O.L. (1989) *The Ecology of Urban Habitats*. Chapman & Hall, London.

Hall, D.A. (1960) *The Soil*. Murray, London.

Hickey, M. and King, C. (2000) *The Cambridge Illustrated Glossary of Botanical Terms*. Cambridge University Press, Cambridge.

Huxley, A. (1981) *The Penguin Encyclopaedia of Gardening*. Allen Lane, London.

International Association for Plant Taxonomey and Nomenclature (1961) *International Code of Nomenclature for Cultivated Plants*. Utrecht.

Krussmann, G. (1985) *Manual of Cultivated Broad-Leaved Trees & Shrubs*, 4 vols. Timber Press, Portland, Oregon.

Leigh, G.J. (2004) *The World's Greatest Fix*. Oxford University Press, Oxford, UK.

Pavord, A. (2005) *The Naming of Names*. Bloomsbury, London.

Shigo, A.L. (1986) *A New Tree Biology*. Shigo and Trees Associates, Durham, New Hampshire.

Watkins, J. and Wright, T. (2007) *The Maintenance and Management of Historic Parks Gardens and Landscapes*. English Heritage Handbook, UK.

Index

Page numbers in **bold** type refer to figures and tables.